ORGANIZING DATA
and
DEALING WITH
UNCERTAINTY

Two units on probability and
statistics adapted from

EXPERIENCES IN MATHEMATICAL IDEAS
Volume 2

NATIONAL COUNCIL OF
TEACHERS OF MATHEMATICS

1906 Association Drive
Reston, Virginia 22091

Second printing 1990

Library of Congress Cataloging in Publication Data:

National Council of Teachers of Mathematics.
Organizing data and dealing with uncertainty.

"Two units on probability and statistics adapted
from Experiences in mathematical ideas, volume 2."
1. Statistics. 2. Probabilities. I. National
Council of Teachers of Mathematics. Experiences in
mathematical ideas. II. Title.
QA276.12.N38 1979   001.4'22   79-9281
ISBN 0-87353-141-8

Printed in the United States of America

Contents

Preface

In its *Overview and Analysis of School Mathematics, Grades K–12,* the National Advisory Committee on Mathematical Education (NACOME) recommended that ideas from probability and statistics should become part of the elementary and secondary school mathematics curricula. The report suggested that although some instruction in probability was beginning to take place, little if any attention was being given to statistical ideas. As a result of a charge from the Board of Directors of the NCTM to examine the NACOME report, the Publications Committee found that few publications of the Council dealt with probability and statistics, particularly for elementary and junior high school students or teachers. However, an existing publication, *Experiences in Mathematical Ideas,* Volume 2, did contain two units devoted to ideas from probability and statistics. The committee decided that these units would be revised as necessary, updated, and reissued as separate instructional units. Thus the materials in the present units, with minor revisions, are derived from chapters 11 and 12 of *Experiences in Mathematical Ideas,* Volume 2.

The initial concept and beginning prototype units of EMI were the result of the NCTM Committee on Mathematics for the Non-College Bound. The entire project was the responsibility of the EMI Executive Committee: Arnold M. Chandler, director; George Immerzeel; and Harold C. Trimble. The EMI writing team included Charles Allen, Beth Baer, Bonnie Brooks, C. William Engel, Lowell Lenke, Jr., Edith Robinson, Carolyn Smith, Donald Wiederanders, and Larry Yarck. Raphael Wagner and Kenneth Travers assisted the Executive Committee with editorial responsibility.

The editors of the present publication have not made any substantive changes in the original materials. The most noticeable change is that all measurement units are now metric.

George Cathcart

Joan Kirkpatrick

Introduction

"Organizing Data" and "Dealing with Uncertainty" are units designed to help teachers provide interesting learning situations involving simple, basic ideas of probability and statistics for children in grades 5 through 8.

As was true of all the EMI materials, these units are presented in such a way that teachers may use them in any organizational pattern—self-contained classrooms and team-teaching or nongraded situations, for example. The units are independent of one another and may be used with any mathematics program in the upper elementary or junior high school grades. They could also be used in general mathematics classes.

Each of the units contains a sequential development of ideas through a series of "experiences." A brief overview of the experiences is outlined at the beginning of each unit to give the teacher a quick preview of what the unit is about. There are five experiences in each unit. Some of these may take up to three days, others only one day.

The experiences in these units should provide the following:

1. *Enjoyment.* Learning should be an enjoyable experience. The experiences in these units involve activities that children will find interesting. It is hoped that through having fun children will be motivated to study mathematics further.

2. *Systematic thinking.* In general our schools do not provide many opportunities for children to develop an organized and systematic way of thinking about things. Some of the experiences in these units do provide an opportunity for systematic thinking by requiring children to collect and organize data and to make predictions based on their data.

3. *Practice.* A considerable amount of practice with, and reinforcement of, basic skills, including computation, is built into the experiences in these units.

4. *Individualization.* Learning is an individual process. There are opportunities in these units for individual children to observe, explore relationships, collect information, hypothesize solutions to problems, test hypotheses, and do independent investigations.

5. *Success.* Success breeds success. These units provide activities that should give every student in your class a successful experience. The activities progress in small steps to the point where a child should be able to grasp the basic idea. In order to achieve success, some children will need to spend more time on some activities than other children.

The description of each experience includes (1) a statement of an objective for the student; (2) a list of materials to be used; (3) a strategy that may be employed by the teacher; (4) reproductions of student activity worksheets; and (5) a general suggestion about evaluation.

The Teaching Package at the end of this booklet consists of student worksheets that teachers can duplicate. The pages are perforated for easy removal. An overhead transparency could be made from the first of the pages for unit 1, experience 4. The Teacher's Guide for unit 2 also contains some supplementary notes on probability.

1

EXPERIENCES WITH

Organizing Data

The purpose of this unit is to help students develop a basic understanding of the nature of descriptive statistics by collecting and organizing data into tables and graphs, then describing the results. Since one can scarcely read a newspaper or magazine without being confronted with statistical information, it seems important that students become aware at an early age of the methods used in presenting the statistics.

Some of these experiences are laboratory-oriented. Through these experiences the students will work both individually and in groups to gather data and analyze them on the basis of information they have obtained from introductory discussions. It will often be necessary to call them together at the end of the class session to give them an opportunity to share their results. Comparing and discussing the different data and descriptions from the same activities will help them understand what statistical data really represent.

OVERVIEW

Each of the five experiences in this unit includes a detailed section called "Teacher Strategy," which is usually one teacher's first-person account of a procedure that proved effective with his or her class.

1

The following brief summary will help you decide whether these experiences meet the needs of your own class and will also give you some indication of the amount of preparation involved.

Student worksheets and other "handout" materials are provided in the Teaching Package. It is assumed that duplicating facilities are available to you.

Experience 1: Data from the Class

Students focus attention on everyday situations from which data can be collected. They are led to see a need for organizing these data in order to get some meaning from what might otherwise be "just a bunch of numbers." They are then given the opportunity to organize some prepared data on their own.

Materials needed: worksheets.

Experience 2: Collecting and Organizing Your Own Data

Students are given the opportunity to use the principles learned in Experience 1 to collect and organize their own data. These data are obtained from gamelike experimental activities. In some of these activities they work as teams, and in others they work individually.

Materials needed: posters, markers, target, paper squares, masking tape, buttons, meterstick, envelopes with paragraphs of prose, dice, group data sheets, worksheets.

Experience 3: Graphing and Interpreting Data

Gamelike activities again serve as a means for the students to collect their own data. This time they work independently. The students are also introduced to bar-graph representations of data, using information they collected in Experience 2. Finally, they construct tables and graphs to aid them in describing the data they collected.

Materials needed: class data from Experience 2, projector transparencies, magazine clippings, construction paper, spinner faces, gummed reinforcements, Popsicle sticks, wire, rulers, worksheets.

Experience 4: More on Interpreting Data

Students investigate relationships between pairs of numbers, the first member from one set and the second member from another set. The introduction is a consideration of the relation between hand lengths and the number of dots counted as a part of Experience 3. They are then shown

instances where there is a distinct relation between pairs of numbers and others where there is no evident relation, so that they will know what to look for in their graphs.

Materials needed: Experience 2 Button Toss materials; Experience 3 worksheets, group report, spinners, and paragraphs; grids; transparency; worksheets.

Experience 5: Culminating Experience

This experience serves as a summary. The students are given a data sheet, some blank grids and tables, and a list of questions. In order for them to answer the questions it is necessary to organize the data into graphs and tables. Thus they will show their competence at using these tools to solve problems.

Materials needed: data sheet, worksheets.

EXPERIENCE 1
Data from the Class

OBJECTIVE

The student should be able to see the need for organizing data and able to record a set of data in a table in an organized manner, using tallies.

MATERIALS

2 worksheets for each student

TEACHER STRATEGY

This experience has a twofold purpose. First, we want the students to see a need for organizing data; secondly, we want them to see how to organize the data. In order to achieve the first goal it is important that data obtained from the students be placed on the board in a random fashion—then they will see that it will be necessary to organize the information that has been collected in order to describe it well.

I like to begin this experience by asking, "Which of you lives the farthest from school?" When I receive an answer, I ask, "How many blocks away from school do you live?" and write the number given in the response on the board under the title "Number of Blocks from School."

"Does anyone live farther away than that? How far?" I put these numbers on the board.

"Now which of you lives closest to school?" I record that number.

"Does anyone live closer? Does anyone live a distance from school that is not listed on the board?" I add these data to the list, then question students in turn until everyone is represented by some number on the board. I make no attempt to enter the numbers systematically; rather, I enter them in order as they are determined.

Now I go back over the list and find out how many students live each of the given numbers of blocks from school. I use tally marks to record this information.

I tell the students that we have gathered the information for a frequency distribution table that tells about the distances the students live from school. I ask them if there is some way we can organize the information in the table so we can readily see the shortest distance, the longest distance, the number of students living more than five blocks away, and so forth. The students suggest listing the distances in order from shortest to longest. After rearranging the data in a table as they suggest, I ask questions like the following to see if the students know how to use the table: "Which distances are the most frequent? How many students walk more than six blocks to school?"

To give the students additional experience in recording data I ask, "How many of you remember what a polygon is?" and have one of the students describe one. If no one remembers, I give them a couple of examples like triangles, squares, or pentagons. Then I ask the students to draw three different polygons and count the number of sides. I ask them to tell me the number of sides of their polygons and make a list on the board like 9, 3, 6, 4, 3, 8, . . . until all the numbers are recorded.

I ask the students for suggestions on organizing the data. They should suggest making a list like the one shown in table 1, which lists an increasing order, using tallies to note the frequency and then showing the totals. From such a table it is easy for the students to decide which polygon was drawn most frequently.

TABLE 1

Number of Sides	Frequency Tallies	Total
3		
4		
5		
6		
7		
8		
9		
10		

Next I ask each student to write down a number between 1 and 10, inclusive, and ask the class to suggest ways to show the data in a table. One like table 2 (p. 7) will probably be suggested. It should be completed.

Then I ask, "Which number was written most frequently? Least frequently? Are there any numbers that weren't written at all? If we did this

TABLE 2

Number	Frequency Tallies	Total
1		
2		
3		
4		
5		
6		
7		
8		
9		
10		

again, do you think the distribution would be the same?" If it seems necessary, I repeat this activity.

With this discussion as background, I pass out copies of the first worksheet and read through the directions with the students. I leave on the board the tables that have been constructed in class, since the students may need some guidance in beginning to make the first tables. As the students complete the first worksheet, I give them a copy of the second one.

When all have completed their worksheets, I use the questions at the end of the sheets as a basis for discussion. Such questions help us to describe the data we collect.

It is important to check to make sure that the students complete their tables correctly.

EVALUATION

The students' responses to the questions will indicate whether they are able to understand what their tables represent. To further check their understanding, have the students gather and organize data from their own class and compare it with the class data from either Activity 1 or Activity 2. (This will also serve as an introduction to Experience 2—Collecting and Organizing Data.)

DATA FROM THE CLASS   UNIT 1, EXPERIENCE 1

Activity 1   Name _____

The students in Classroom 204 turned in a record of their birthdays. Here they are:

Jan. 20	Oct. 20	Feb. 14	Aug. 16	Feb. 12
Aug. 15	Nov. 12	Aug. 18	Jan. 2	Dec. 20
Apr. 12	Apr. 15	Mar. 20	Aug. 14	Jan. 15
Sept. 10	Jan. 9	Oct. 10	June 9	Oct. 14
Feb. 15	Sept. 22	June 9	Nov. 11	Jan. 20

Use this information to complete the frequency tables below. Then answer the questions.

Month	Tally	Frequency
Jan.	ЖН	5
Feb.	|||	3
Mar.			1
Apr.	||	2
May		0
June	||	2
July		0
Aug.	||||	4
Sept.	||	2
Oct.	|||	3
Nov.	||	2
Dec.			1

Day	Tally	Frequency	Day	Tally	Frequency	Day	Tally	Frequency
1		0	11			1	21		0
2			1	12					3	22			1
3		0	13		0	23		0
4		0	14					3	24		0
5		0	15						4	25		0
6		0	16			1	26		0
7		0	17		0	27		0
8		0	18			1	28		0
9					3	19		0	29		0
10	||	2	20	ЖН	5	30		0

What month has the most births? _Jan._ How many? _5_

Do any months have no births? _Yes_ If so, which one(s)? _May, July_

How many months have at least 3 births? _4_

How many have less than 2? _4_

On what day of the month did most births occur? _15_

On what days were there no births? _1, 3, 4, 5,_
_6, 7, 8, 13, 17, 19, 21, 23, 24,_
_25, 26, 27, 28, 29, 30_

DATA FROM THE CLASS   UNIT 1, EXPERIENCE 1

Activity 2   Name _____

The data below came from a survey of the number of brothers and sisters of students in a certain class. "B" means brothers and "S" means sisters, so that 2B 3S means 2 brothers and 3 sisters.

2B 3S	2B 3S	3B 1S	1B 3S
3B 2S	0B 0S	3B 3S	1B 4S
1B 0S	2B 0S	2B 2S	3B 4S
4B 2S	0B 2S	0B 0S	1B 4S
1B 1S	2B 5S	3B 1S	2B 3S
3B 1S	4B 0S	4B 1S	0B 4S
2B 2S	0B 2S	2B 1S	6B 1S
1B 3S	2B 0S	2B 2S	1B 2S

Complete the frequency tables below and then answer the questions that follow.

Number of Brothers	Tally	Frequency
0	̶H̶H̶	5
1	̶H̶H̶ II	7
2	̶H̶H̶ ̶H̶H̶	10
3	̶H̶H̶ I	6
4	III	3
5		0
6	I	1

Number of Sisters	Tally	Frequency
0	̶H̶H̶ I	6
1	̶H̶H̶ II	7
2	̶H̶H̶ III	8
3	̶H̶H̶ I	6
4	IIII	4
5	I	1

Total Number of Both	Tally	Frequency
0	II	2
1	I	1
2	̶H̶H̶	5
3	II	2
4	̶H̶H̶ ̶H̶H̶	10
5	̶H̶H̶ II	7
6	II	2
7	III	3

The greatest number of brothers anyone has is ___6___ .

The greatest number of sisters is ___5___ .

The least number of brothers is ___0___ .

The least number of sisters is ___0___ .

What is the most frequent number of brothers? ___2___

Of sisters? ___2___

Of brothers and sisters? ___4___

– 2 –

EXPERIENCE 2
Collecting and Organizing Your Own Data

OBJECTIVE

The student should be able to collect data from experimental situations and organize them in a meaningful fashion in the tables provided.

MATERIALS

4 posters with class data sheets
Markers
Station and activity materials as listed below

About one-fourth of the students in the class will be working on each activity at one time. They work in groups of three in the first two activities and individually in the last two. Table 3 indicates the number of sets of materials (apart from worksheets) needed for various class sizes.

TABLE 3

Size	1–4	5–8	9–12	13–16	17–20	21–24
Activity 1	1	1	1	2	2	2
Activity 2	1	1	1	2	2	2
Activity 3	1	2	3	4	5	6
Activity 4	1	2	3	4	5	6

Activity 1, Station A—"Skydiver over Target"

1 chair
Target
5 squares of construction paper, each 5 cm × 5 cm
1 individual worksheet for each student
1 worksheet for each group

The target is made from a large square of cardboard or paper with five concentric circles that have radii of 5, 15, 25, 35, and 45 cm as shown in figure 1 on the following page. It should be taped to the floor if necessary to

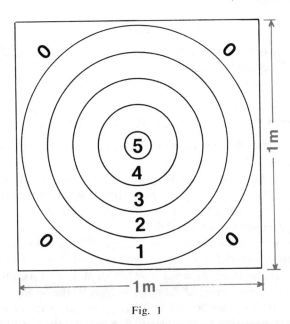

Fig. 1

keep it flat and in place. The student stands on the chair next to the target and drops the five squares over the target, one at a time, from shoulder height.

Activity 2, Station B—"Button Toss"

Circular cardboard disc (3 cm diameter) or other marker for target
Masking tape
5 matching buttons
Meterstick or tape measure
1 individual worksheet for each student
1 worksheet for each group

Tape the disc to the floor or make a mark on the floor. With chalk, string, or masking tape make a line about two meters from the target. The student stands behind the line and tosses five buttons, one at a time, at the target.

Activity 3—"Vowel Count"

1 envelope for each student, containing a paragraph clipped from a
   magazine or newspaper
1 worksheet for each student

The magazine or newspaper clippings may be mounted on construction paper for greater durability. Directions for this activity are given on the student worksheet.

Activity 4—"Horseshoe Game"

1 button or other small object as a marker for each student
1 die for each student
2 worksheets for each student, the first of which gives the directions

TEACHER STRATEGY

In this experience the students collect and organize their own data, both individually and in groups of three.

Divide the class into four main groups, then divide each of these groups into teams of three. Have each team select one chairman for Activity 1 and another for Activity 2. It is the responsibility of the chairman to record the data for the team activity.

It will be necessary for you to control the time spent in the four activities so that there is no conflict about the time spent on Skydiver over Target and Button Toss; the time spent on Vowel Count and Horseshoe Game can vary, since they are done individually, so long as the total of the two is the same as for the two team activities. A fifteen-minute warning is suggested so that students have time to complete their work and the chairmen for Activities 1 and 2 will have time to report the team results.

In Skydiver over Target, Activity 1, each of the three players takes three turns at each position on the team. Team Member 1 takes five of the squares, stands on a chair next to the target, and drops the squares over the target, one at a time, from shoulder height. Team Member 2 announces the spaces where the squares land, counting a square as landing in the zone with the higher number if it lands equally in two target zones. Team Member 3 enters tallies in the Individual Data Sheet for Team Member 1. After three turns apiece, the players find their individual scores for each of the three trials. Together, the players figure out the remaining data for the team data sheet; then the chairman enters the scores for his or her team on the poster for the class. The team data sheet is retained for future class discussion.

Button Toss, Activity 2, is conducted in much the same way. Each player has a turn at each position on the team—or two turns, if time allows. Team Member 1 tosses five buttons, one at a time, from behind a line placed about two meters from the target. Team Member 2 measures the distance in centimeters that each button lands from the target. Team Member 3 records this information on Team Member 1's individual data sheet. At the warning signal, the players complete the individual and team data sheets, and the chairman enters the information on the poster for the class.

The results of these activities are more interesting to discuss if the students are able to *see* the compilation of data obtained from all the students. An easy way to show this compilation is to have large pieces of poster paper taped around the room for a summary of the reports, as shown in figure 2 for the team scoring data reports of the Button Toss game. Team captains and individuals record their information, as already mentioned, at the completion of each activity. These reports will be used as material for class discussion and as a kickoff for the use of graphs, in the next experience, to represent data.

When the students have completed all the activities ask them to describe their results to the class. Some questions like the following will help them make this description.

What was the least number of tosses required in any horseshoe game? What was the greatest number? Which number of tosses appeared most frequently?

What vowel was used most frequently in your sentences? What vowel was used the least?

Do your results agree with those compiled by the entire class?

The class data sheets should be saved, since some of the data are used in Experience 3.

BUTTON TOSS
Class Data Sheet

Scores	Frequency								Total Frequency
	Team 1	Team 2	Team 3	Team 4	Team 5	Team 6	Team 7	Team 8	
0–4									
5–9									
10–14									
15–19									
20–24									
25–29									
30–34									
35–39									
40–44									
45–49									
50									

Fig. 2

EVALUATION

The discussion and the completed worksheets will indicate how well the students can collect and organize data.

COLLECTING AND ORGANIZING
YOUR OWN DATA   UNIT 1, EXPERIENCE 2

Activity 1, Station A   Name _____

SKYDIVER OVER TARGET

Stand on a chair next to the target. Drop the five squares, one at a time, onto the target from shoulder height. Do this three times.

Individual Data Sheet

Target Area	Trial 1		Trial 2		Trial 3	
	Tally	Score	Tally	Score	Tally	Score
5						
4						
3						
2						
1						
0						
Score Totals						

To obtain the score, multiply the number of the target area by the number of squares that landed there. For example, if the frequency for Target Area 5 is 2, the score to be entered for that area would be 10.

Score for Trial 1 _____

Score for Trial 2 _____

Score for Trial 3 _____

COLLECTING AND ORGANIZING
YOUR OWN DATA   UNIT 1, EXPERIENCE 2

Activity 1, Station A   Name _____

                                     Name _____

                                     Name _____

SKYDIVER OVER TARGET

Team Data Sheet

Summary of Scores

Name	Trial 1	Trial 2	Trial 3

Distribution of Scores for the Team

Scores	Tally	Frequency
0–4		
5–9		
10–14		
15–19		
21–25		

COLLECTING AND ORGANIZING
YOUR OWN DATA   UNIT 1, EXPERIENCE 2

Activity 2, Station B   Name _____

BUTTON TOSS

Stand behind the line and toss the five buttons, one at a time, at the target. Measure the distance each button lands from the target and enter the tallies under Trial 1. Repeat for Trials 2 and 3. To obtain your score multiply the number of tallies by the points for each tally.

Individual Data Sheet

Points for Each Tally	Distance from Target in Cm	Trial 1		Trial 2		Trial 3	
		Tally	Score	Tally	Score	Tally	Score
10	0–10						
9	11–20						
8	21–30						
7	31–40						
6	41–50						
5	51–60						
4	61–70						
3	71–80						
2	81–90						
1	91–100						
0	101 or more						
	Total Scores						

Score for Trial 1 _____

Score for Trial 2 _____

Score for Trial 3 _____

– 5 –

COLLECTING AND ORGANIZING
YOUR OWN DATA

UNIT 1, EXPERIENCE 2

Activity 2, Station B

Name _____

Name _____

Name _____

BUTTON TOSS

Team Data Sheet

Summary of Scores

Team	Trial 1	Trial 2	Trial 3
Team Member 1			
Team Member 2			
Team Member 3			

Distribution of Scores for the Team

Scores	Tally	Frequency
0–4		
5–9		
10–14		
15–19		
20–24		
25–29		
30–34		
35–39		
40–44		
45–49		
50		

– 6 –

COLLECTING AND ORGANIZING
YOUR OWN DATA

UNIT 1, EXPERIENCE 2

Activity 3

Name _____

VOWEL COUNT

From your English classes, you are already familiar with the vowels *a, e, i, o*, and *u*. I'll bet you never expected to use them in math. Actually, all we're trying to do is find out which ones are used the most.

Take the clipping out of your "Vowel Count" envelope.

Look at the first word. Suppose it is "today." For that word you would put a tally in the *o* column and another tally in the *a* column. Do this for every word in the sentences you have. When you have finished, find the total number of tallies for each vowel and place that number in the last column. Then place your results in the proper place on the sheet for the class report.

Vowel	Tally	Frequency
a		
e		
i		
o		
u		

COLLECTING AND ORGANIZING   UNIT 1, EXPERIENCE 2
YOUR OWN DATA

Activity 4   Name _____

HORSESHOE GAME

The object of this game is to move your marker from start to finish in as few moves as possible.

Place your marker at the starting line. Toss the die. Move the marker forward along the horseshoe as many spaces as the top of the die indicates, and record a tally mark in the Tally of Tosses column on your Individual Data Sheet for this activity. Continue tossing the die, moving forward the indicated number of spaces, and recording a tally mark for each toss.

You must cross the finish line on an exact count. If your die toss results in a number more than the number of spaces that remain you must keep your marker where it is, record a tally for the toss, and toss again (with a tally each time) until you get a number that you can use.

You may continue to play the game until the warning signal is given by your teacher. Then complete the frequency table on the Individual Data Sheet and record your data on the table for the class.

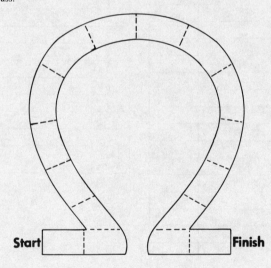

20 ORGANIZING DATA

COLLECTING AND ORGANIZING
YOUR OWN DATA

UNIT 1, EXPERIENCE 2

Activity 4   Name _____

HORSESHOE GAME

Individual Data Sheet

Trial	Tally of Tosses	Frequency
1		
2		
3		
4		
5		
6		
7		
8		
9		
10		
11		
12		
13		
14		
15		
16		
17		
18		
19		
20		

Summary of Trials

Number of Tosses	Tally	Frequency
3		
4		
5		
6		
7		
8		
9		
10		
11		
12		
13		
14 or More		

- 9 -

EXPERIENCE 3
Graphing and Interpreting Data

OBJECTIVE

The students should be able to make a representation in bar-graph form of the data they have collected and make meaningful statements to describe the data, indicating an understanding of range and average.

MATERIALS

Class data from Experience 2

Projector transparency showing bar graph from Button Toss, Experience 2

Station materials (see below)

9 worksheets for each student

Duplicate stations will probably be needed. The number depends on the size of your class.

Station A, Letter Count. A short paragraph from a magazine or newspaper, enclosed in an envelope, is needed for each student expected to work at the station at any one time. The clipping may be mounted on construction paper for reinforcement. Extra envelopes might be supplied so that a student who wishes to analyze a second paragraph and collect additional data would be able to do so. All clippings at any one station should be from the same source.

Station B, Spinner Sums. For this activity you should prepare enough spinners to provide one for each student working on this activity at any one time.

Spinners

Mount the faces, duplicated from the Teaching Package, on medium-weight cardboard. Make a small hole in the center of each spinner and add about 10 gummed reinforcements around the hole to serve as a washer. Using a compass point, punch a hole in the midpoint of a Popsicle stick, then attach it to the disc with a thumbtack. Identify the pointing end of the Popsicle stick in some way—with poster paint, crayon, or nail polish, for example.

Figure 3 pictures the completed spinner.

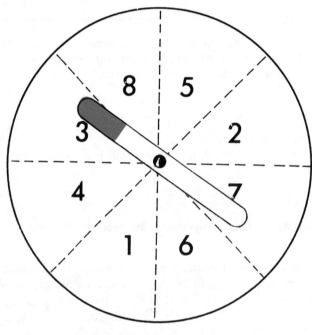

Fig. 3

Station C, Dot Count. Only rulers are needed for this station, in addition to the worksheet. There should be a ruler for each student.

TEACHER STRATEGY

I like to begin this experience by considering the class results for the Button Toss game of Experience 2. I have at hand a record of the total frequencies of the scores and an overhead projector transparency of a bar-graph representation of this information. (If no projector is available, a large-scale graph on manila paper could be substituted for the transparency.) I show the graph to the students along with the frequency table and try to elicit the observation from students that graphs are useful because they picture the numerical facts in a frequency table and help us to summarize the data relationships clearly and quickly.

I ask students to tell me how to construct a bar graph such as that shown on the transparency, using the data in a table. In the discussion that follows I emphasize that it is important to have a column for each entry to that table and add that columns should be of equal width, each vertical space should represent the same number of units, and so forth. Referring to the transparency, I ask questions such as "What total score was achieved most often?" and "What were the lowest and highest scores?"

I also ask them, "From the graph, what appears to be the 'average' score?" After they have answered this question, we discuss adding up all scores and dividing; then the students use this method to calculate the average for this activity. Since we want whole number scores, we round the average to the nearest whole number and round up in case of an average involving one-half. The average is usually quite close to that previously chosen by the students.

This demonstration should show them that by a careful examination of a graph of data they can come up with a close "average" for the data.

Now I show the students the class data sheet from the Skydiver over Target game, showing the scores, together with frequencies. I give the students the worksheet for Activity 1 and direct them to construct the graphs for this set of data. We discuss a scale for the vertical columns and determine a scale according to the amount of data collected. After they have graphed the data in the first column (frequency of landings in the zero target area), I have them continue individually. They should have little trouble completing the graph.

When the graph has been completed, I have them describe their results in a way that answers questions like those they answered in connection with Button Toss. The descriptions should indicate the range, most frequent scores, average score, and so on.

For the next three activities I divide the class into three groups, so that group data reports can be made; but the students work individually on the worksheets for these activities and for Activity 5.

The group data report forms should be posted in three sections of the room. They may be made from large pieces of manila paper or poster paper, and they should be fashioned like the one indicated for Spinner Sums in figure 4.

You will need to allow time during the day for students to obtain the information from the group data sheets. You can either allow time in class for students to complete the graphs and descriptions or assign this to be done as homework. Perhaps it would be better to have this done during class time, for then you would know it was their work and you could better evaluate their understanding.

SPINNER SUMS

Sum	Tally			Total Frequency
	Group 1	Group 2	Group 3	
2				
.				
.				
.				
16				

Fig. 4

After the students have had an opportunity to complete all the worksheets, discuss the questions they were to answer. Have someone from each group put a sample graph on the board so students can compare results.

In talking about the Letter Count activity, let the students compare the numbers of letters in words from different sources. Ask why there are so many words of four and five letters in their paragraphs. Have them think of the words they use. You might have a second-grade reader and a college textbook to show the students, letting them see how the lengths of the words compare.

In connection with the Spinner Sum activity, discuss some questions like the following:

Can you think of a reason why certain sums hardly ever occurred?
Is it just by chance?
Why do some appear so often?
What are the ways we can get certain sums—for instance, 8 or 9 and 2 or 16?

Save the Dot Count data for Experience 4.

EVALUATION

The responses of the students to the discussion questions and their work on the worksheets should give you a very good idea of their competence in describing data they have collected.

GRAPHING AND INTERPRETING DATA   UNIT 1, EXPERIENCE 3

Activity 1   Name _____

Follow the directions given in class to construct a bar graph to represent the class data from the Skydiver over Target game.

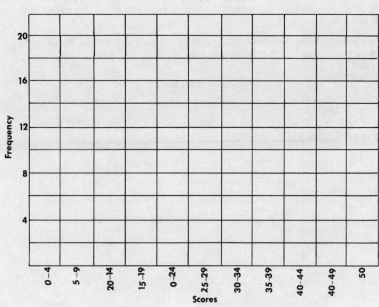

Frequency of Scores

GRAPHING AND INTERPRETING DATA   UNIT 1, EXPERIENCE 3

Activity 2, Station A   Name _____

LETTER COUNT

Take a paragraph out of one of the Letter Count envelopes.

Look at the first word. Suppose it is the word *when*. *When* is a 4-letter word, so you should put a tally mark after the 4 in the Tally column of the table below. Use hyphenated words as one word.

When you have tallied the number of letters for each word in your paragraph, find the total number of tallies for each number of letters and put that number in the last column.

Individual Data Sheet

Number of Letters in Word	Tally	Frequency
1		
2		
3		
4		
5		
6		
7		
8		
9 or More		

Enter these results in your group data report, then go on to the next page.

[Continued]

GRAPHING AND INTERPRETING DATA   UNIT 1, EXPERIENCE 3

Activity 2—*Continued*   Name _____

Make a graph of the information in your Individual Data Sheet.

Individual Data

When your group data sheet has been completed, make a graph of the information. No form is provided.

GRAPHING AND INTERPRETING DATA   UNIT 1, EXPERIENCE 3

Activity 3, Station B   Name _____

SPINNER SUMS

Pick up a spinner from Station B and take it to your desk. Keep it flat on your desk and spin the pointer once. Look at the number that the pointer indicates. If a pointer stops on a line, spin it again. Then spin it again to get a second number.

Find the sum of the numbers. Write the sum after Trial 1, below. Continue spinning the pointer and finding the sums until you have filled in all the spaces in the charts below.

Trial	Sum	Trial	Sum	Trial	Sum
1		17		33	
2		18		34	
3		19		35	
4		20		36	
5		21		37	
6		22		38	
7		23		39	
8		24		40	
9		25		41	
10		26		42	
11		27		43	
12		28		44	
13		29		45	
14		30		46	
15		31		47	
16		32		48	

[*Continued*]

GRAPHING AND INTERPRETING DATA   UNIT 1, EXPERIENCE 3

Activity 3—*Continued*   Name _____

Look at your data on the preceding page. What is the smallest sum you got? Put it on the first line below. What is the largest sum? Put it on the bottom line. Fill the other sums in, in order from smallest to largest. Record tally marks after them to indicate the number of times you obtained each sum.

Individual Data Sheet

Sum	Tally	Frequency
Smallest Sum =		
Largest Sum =		

Record the above data on your Group Data Sheet.

[*Continued*]

GRAPHING AND INTERPRETING DATA   UNIT 1, EXPERIENCE 3

Activity 3—*Continued*   Name _____

Complete this chart just as you did when you organized your own data except that this time
you should use the data reported by your group on the group data report form. When you
have completed the chart, use the information to construct a graph in the grid provided
on the next page.

Group Data Sheet

Sum	Frequency
2	
3	
4	
5	
6	
7	
8	
9	
10	
11	
12	
13	
14	
15	
16	

[Continued]

GRAPHING AND INTERPRETING DATA   UNIT 1, EXPERIENCE 3

Activity 3—*Continued*   Name _____

Graph of Group Data

GRAPHING AND INTERPRETING DATA   UNIT 1, EXPERIENCE 3

Activity 4   Name _____

DOT COUNT

Hold your hand over the bottom part of this sheet. Close your eyes. Drop your hand, flat, on the sheet. Trace around your hand with a pencil. Stop at the base of your palm. Then count the number of dots in the area your hand covered. To make counting easier, do it by areas. For example, count the dots in each of the fingers and then divide the palm into regions and count them.

After you have counted the dots, measure the distance from the tip of your longest finger to the base of your palm as you have traced your hand on the paper. Also measure the length of one of your shoes. Enter this information in the spaces provided below and on your group report sheet.

Number of dots inside sketch of hand _____

Length of hand to nearest cm _____ Length of shoe to nearest cm _____

GRAPHING AND INTERPRETING DATA   UNIT 1, EXPERIENCE 3

Activity 5   Name _____

Study the tables and graphs you have made in the activities in this experience, then use them to help you answer the following questions:

1. How many different sums did you find, using the two spinners? _____

2. Did you get some of the sums more often than others? _____

3. What sum(s) did you get most often? _____

4. What sum(s) did you get least often? _____

5. Did you ever get a sum of 20? _____ If not, why not? _____

    _____

6. How many letters were there in the shortest word you had? _____

7. How many letters were there in the longest word you had? _____

8. Do you think that you might have had different results if you had used a different magazine? _____

9. In general most words in the paragraph are   3   4   5   6   letters in length. Circle the number(s) that apply.

10. Do you think anyone else's Dot Count will be exactly the same as yours? _____

11. Do you think your Dot Count will be high, low, or in the middle of the data collected by the entire class? _____

12. Do you think that the size of your hand makes any difference in the number of your Dot Count? _____

EXPERIENCE 4
More on Interpreting Data

OBJECTIVE

The student should be able to determine whether there is any apparent relation between different types of data, using graphical methods.

MATERIALS

Experience 2 Button Toss materials
Experience 3 worksheets and group reports
Experience 3 spinners and paragraphs
3 grids
1 transparency (from Teaching Package) showing 3 graphs
8 worksheets for each student

TEACHER STRATEGY

I like to introduce this experience by pointing out that there are other ways of representing and describing data in addition to the frequency tables and graphs we have been using. One way is to construct a graph containing two different types of data, like the one in the Teaching Package comparing heights and weights of the boys in a classroom, and look for a pattern. I project this completed graph, explain how the data were placed on the grid, and ask the students if they notice any pattern in this set of points. Someone should see that they form a path in an upward motion from bottom left to upper right. Then see if anyone can explain the meaning of the pattern. The students may suggest that as the height increases, so does the weight; or they may say that the points lie near a line that is higher on the right.

Then I project another part of the same transparency. This one pairs the height of each student in a mathematics class with his grade.

I ask the students to describe any pattern that they see. When they observe that there seems to be no pattern and that the dots are quite well scattered over the entire grid, I tell them that there is no simple relation between the heights of students in this class and their grades.

Now I remind the students that they collected some interesting data in Experience 3, ask what kind of graph they would get if they plotted the Dot Count data, and project the grid for this purpose. Several may guess that the number of dots will increase with the length of the hand, giving an upward movement as in the graph for heights and weights. I ask each student to report his Dot Count data and plot the dots on the transparency. When all the points have been plotted I ask the students to describe the relationship, if any. (There may be one, but my own experience has been that because of the random scattering of the dots there is no clear pattern.)

At this point I have the students return to the three groups established during Experience 3. Each group is given a grid marked "Foot Sizes" on the vertical axis and "Hand Sizes" on the horizontal axis. I direct the students to look over their group report sheet for the data, decide on the intervals to be listed in both margins, and plot the data. After this has been done, I lead a brief discussion of the results of each of the three groups.

For the worksheet activities, I regroup the students in teams of three. Both Spinner Relations and Word Lengths and Number of Vowels are essentially individual projects, but the teams should stay together and share the materials at each station. The setup for Button Toss is like that in Experience 2 except that the students make their tosses standing behind arcs of circles with center at the target and radii of 0.5, 1.0, 1.5, 2.0, 2.5, and 3.0 meters. Each student makes five tosses from behind each of the six arcs; while one tosses the buttons a second measures the distances and a third records the data. If time permits, each student gets a second turn and the tallies of distances can be recorded in column 2 of the Individual Data Sheet.

At the conclusion of these activities I have each student show one of the graphs and discuss the relationship he or she found. I have other students point out any differences in their results.

EVALUATION

The discussion and individual worksheets should show that each student was able to graph data from various sources and discover relationships by graphing the data.

MORE ON INTERPRETING DATA   UNIT 1, EXPERIENCE 4

Activity 1   Name _____

BUTTON TOSS

Complete the Individual Data charts, then have one member of your group complete the Group Data Sheet on the next page.

Individual Data

Line A (0.5 m)

Distance in Cm	Tally	
	1	2
0–10		
11–20		
21–30		
31–40		
41–50		
51–60		
61–70		
71–80		
81–90		
91–100		
101 or More		

Scores ____ ____

Line B (1.0 m)

Distance in Cm	Tally	
	1	2
0–10		
11–20		
21–30		
31–40		
41–50		
51–60		
61–70		
71–80		
81–90		
91–100		
101 or More		

Scores ____ ____

Line C (1.5 m)

Distance in Cm	Tally	
	1	2
0–10		
11–20		
21–30		
31–40		
41–50		
51–60		
61–70		
71–80		
81–90		
91–100		
101. or More		

Scores ____ ____

Score Key

Distance in Cm	Points	Distance in Cm	Points	Distance in Cm	Points
0–10	10	41–50	6	81–90	2
11–20	9	51–60	5	91–100	1
21–30	8	61–70	4	101 or More	0
31–40	7	71–80	3		

[Continued]

MORE ON INTERPRETING DATA   UNIT 1, EXPERIENCE 4

Activity 1—*Continued*   Name _____

Line D (2.0 m)		
Distance in Cm	Tally	
	1	2
0–10		
11–20		
21–30		
31–40		
41–50		
51–60		
61–70		
71–80		
81–90		
91–100		
101 or More		

Scores _____ _____

Line E (2.5 m)		
Distance in Cm	Tally	
	1	2
0–10		
11–20		
21–30		
31–40		
41–50		
51–60		
61–70		
71–80		
81–90		
91–100		
101 or More		

Scores _____ _____

Line F (3.0 m)		
Distance in Cm	Tally	
	1	2
0–10		
11–20		
21–30		
31–40		
41–50		
51–60		
61–70		
71–80		
81–90		
91–100		
101 or More		

Scores _____ _____

One of your group members should make a summary of the scores in the report below.

Group Data Sheet

Line A	Line B	Line C	Line D	Line E	Line F

[Continued]

– 22 –

MORE ON INTERPRETING DATA   UNIT 1, EXPERIENCE 4

Activity 1—*Continued*   Name _____

Complete the chart and the graph below.

Group Results

Scores	
A Line (0.5 m)	
B Line (1.0 m)	
C Line (1.5 m)	
D Line (2.0 m)	
E Line (2.5 m)	
F Line (3.0 m)	

Scores:
46–50
41–45
36–40
31–35
26–30
21–25
16–20
11–15
6–10
0–5

Distances in Meters: 0.5 1.0 1.5 2.0 2.5 3.0

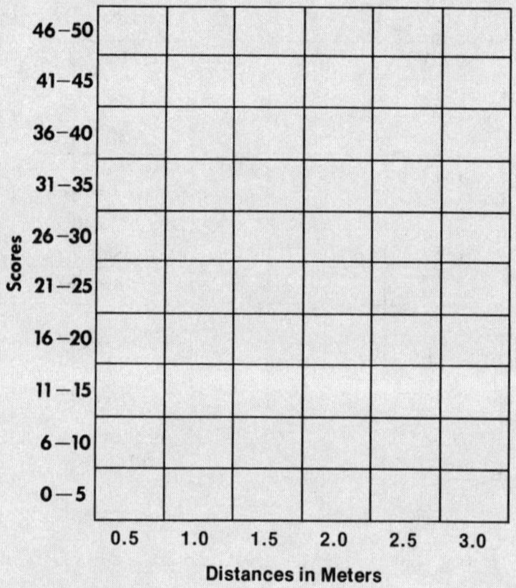

According to your group's graph, did there appear to be any relation between the distances you stood from the target line and the scores you made? _____

If so, what relation did you notice? _____

_____

From which line were the highest scores made? _____

From which line were the lowest scores made? _____

MORE ON INTERPRETING DATA   UNIT 1, EXPERIENCE 4

Activity 2   Name _____

SPINNER RELATIONS

Place your spinner flat on your desk. Spin the pointer once and record the number in the first column of table 1. Spin it again and record the number in the second column of the same table. Then, in table 2, record the number for Spin 1 and the sum of the numbers on the two spins.

Continue doing this, recording the results in the two tables, until you have completely filled in all the spaces. Then go on to the next page.

Table 1

Numbers on Spinner		Numbers on Spinner	
Spin 1	Spin 2	Spin 1	Spin 2

Table 2

Spin 1 and Sums		Spin 1 and Sums	
Spin 1	1 and 2	Spin 1	1 and 2

[Continued]

- 24 -

MORE ON INTERPRETING DATA   UNIT 1, EXPERIENCE 4

Activity 2—*Continued*   Name _____

   Use the pairs of numbers that you collected for table 1 of Spinner Relations to construct the graph below.

   For each pair of numbers there should be a dot on the graph. For example: If the first pair of numbers was a 5 on Spin 1 and a 2 on Spin 2, you should place a dot, as shown, in the space where the fifth vertical column and the second horizontal column intersect.

   Answer the questions below, then go on to the next page.

   Does your graph seem to indicate that there is any relation between the numbers on

Spin 1 and Spin 2? _____

   If so, what relation do you notice? _____

_____

   Did you expect to find a relation? _____

                                                [*Continued*]

MORE ON INTERPRETING DATA   UNIT 1, EXPERIENCE 4

Activity 2—*Continued*   Name _____

Use the same procedure as before, but this time use the data from table 2 and look for a relation between the numbers on Spin 1 and the sum of the numbers on Spins 1 and 2.

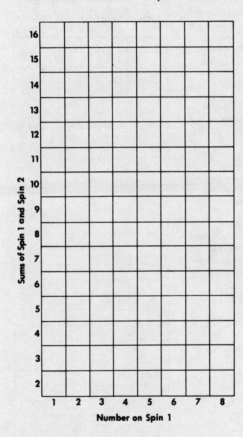

Does your graph seem to indicate that there is any relation between the numbers on Spin 1 and the sums of the numbers of Spins 1 and 2? _____

If it does, what relation do you notice? _____
_____
_____

Did you expect to find a relation? _____

Do you think that there would be a relation between the numbers on Spin 2 and the sums of the numbers of Spins 1 and 2? _____

– 26 –

MORE ON INTERPRETING DATA   UNIT 1, EXPERIENCE 4

Activity 3   Name _____

WORD LENGTHS AND NUMBER OF VOWELS

Individual Data

How do you think the number of vowels a word has compares with the number of letters it has? The longer the word the more vowels? How about the words *a* and *check?* Both words have only 1 vowel, but one has 1 letter and the other has 5. Perhaps these are just exceptions?

Take a paragraph from the envelope. Look at the first word. Put the number of letters it has in the top space of the first column and the number of vowels in the top space of the second column. Make one entry in the chart for each word in your paragraph.

Make a graph of the data in the chart on the grid provided on the next page. There should be one dot in your graph for each pair of numbers that appears in your chart.

When the graph is completed, look to see if there seems to be any relation between the number of letters in the word and the number of vowels it has. Use the space provided under the graph to describe what you find.

Letters in Word	Vowels in Word	Letters in Word	Vowels in Word

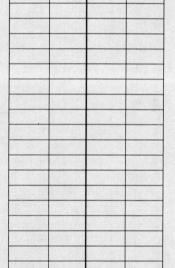

[*Continued*]

- 27 -

MORE ON INTERPRETING DATA   UNIT 1, EXPERIENCE 4

Activity 3—*Continued*   Name _____

WORD LENGTHS AND NUMBER OF VOWELS

Relation Graph

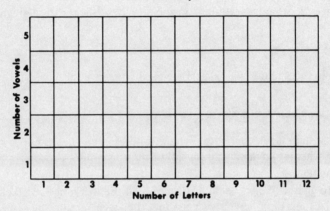

Describe the relation that seems to be shown by the points on your graph. If there seems to be no relation, say so and explain why. _____

_____

_____

_____

EXPERIENCE 5
Culminating Experience

OBJECTIVE

The students should be able to use the methods developed in the previous experiences to organize, tabulate, graph, and describe the data given them.

MATERIALS

1 data sheet and 5 worksheets for each student

TEACHER STRATEGY

This is a summary experience. There is nothing new in it, but it is aimed at finding whether students can put all of their techniques to use in meeting a new situation.

I like to give students the data sheet and all the worksheets and ask them to organize their data in whatever manner they wish so that they can answer the questions on the Summary Sheet (Activity 2). If a student needs help, I ask him or her to make a start on the first chart and then, if help is still needed, to ask me to check the start. Most students are then able to do the other graphs on their own.

After the students have completed their work, I have different ones place a table on the board and then discuss the description.

We use the summary sheets as a basis for discussion.

EVALUATION

The graphs and tables should give quite a complete picture of the students' understanding.

Through class discussion of answers on the Summary Sheet, you should be able to see if the students are able to interpret what they have tabulated. This class-sharing time serves as an additional learning experience for students who have been still unsure of the ideas presented in this unit.

CULMINATING EXPERIENCE   UNIT 1, EXPERIENCE 5

Football Roster
STATE UNIVERSITY

Number	Height in Cm	Weight in Kg	Grade Average
15	184	77	A
70	183	92	B
72	185	95	B
67	181	92	C
43	180	88	C
61	181	92	B
78	183	99	A
74	180	91	B
76	183	97	C
38	177	79	B
51	188	86	C
28	183	81	A
27	178	78	D
30	175	79	B
54	175	92	B
53	182	90	C
25	179	76	B
75	182	92	A
60	179	88	D
20	182	77	A

Number	Height in Cm	Weight in Kg	Grade Average
47	183	90	B
24	181	78	C
22	175	74	B
19	181	78	B
42	178	85	A
34	186	79	D
36	178	88	D
18	181	79	B
65	185	95	C
31	178	75	B
62	173	90	C
33	180	82	A
83	188	104	A
49	180	82	B
88	190	108	B
58	179	86	C

CULMINATING EXPERIENCE   UNIT 1, EXPERIENCE 5

Activity 1   Name _____

Using data from the Football Roster sheet, complete the two graphs below and a third one, at the top of the next page.

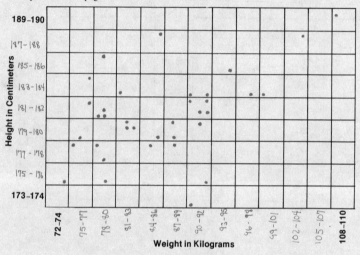

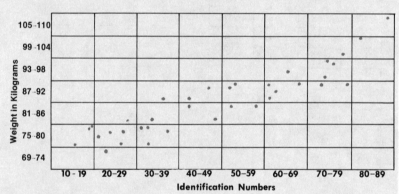

[*Continued*]

CULMINATING EXPERIENCE   UNIT 1, EXPERIENCE 5

Activity 1—*Continued*   Name _____

Grades and Weight

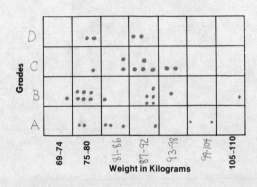

Use the grid at the right
to make any other
comparison that you
would like to consider.

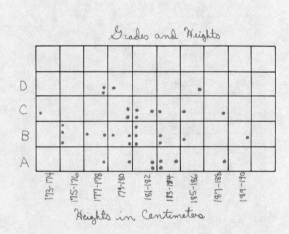

[*Continued*]

48 **ORGANIZING DATA**

CULMINATING EXPERIENCE   UNIT 1, EXPERIENCE 5

Activity 1—*Continued*   Name _____

Using the data sheet again, complete the frequency tables below.

	Height	Tally	Frequency
Shortest =	173–174	I	1
	175-176	III	3
	177-178	IIII	5
	179-180	IIIIII	7
	181-182	IIIIIII	8
	183-184	IIIII I	6
	185-186	III	3
	187-188	II	2
Tallest =	189–190	I	1

Weight	Tally	Frequency
72–74	I	1
75–77	IIII	4
78–80	IIIII II	7
81–83	III	3
84-86	III	3
87-89	III	3
90-92	IIIII IIII	9
93-95	II	2
96-98	I	1
99-101	I	1
102-104	I	1
105-107		0
108–110	I	1

Grade	Tally	Frequency
A	IIIII IIII	8
B	IIIII IIIII IIIII	15
C	IIIII IIII	9
D	IIII	4

[*Continued*]

– 32 –

CULMINATING EXPERIENCE   UNIT 1, EXPERIENCE 5

Activity 1—*Continued*   Name _____

Using the frequency tables you have just made, complete the graphs below.

Heights in Centimeters

Grades

Weights in Kilograms

– 33 –

CULMINATING EXPERIENCE   UNIT 1, EXPERIENCE 5

Activity 2   Name _____

Summary Sheet

1. What is the height of the tallest player? _190 cm_

2. What is the height of the shortest player? _173 cm_

3. The most common height is _181-182 cm_

4. Which player is the lightest? _#22_

5. Which player is the heaviest? _#88_

6. From your frequency graph, what would you consider to be an average weight for the team? _86-87 kilograms_

7. Are these players good students? _Yes_

8. How many have an A average? _8_

9. More players have a(n) _B_ average than any other.

10. The average for the team is in which range? Circle one.

    A–B   (B–C)   C–D   D–F

11. What relations did you find?

    a) Does there appear to be any relation between height and weight? _Yes_

    If so, state some specific relation. _Weight increases with height._

    b) Does there appear to be any relation between the players' numbers and their weight? _Yes_ If so, state some specific relation. _The heavier players have higher numbers._ (Customary for linemen.)

    c) Is there any relation between grades and weight? _No_

    If so, state some specific relation. _Mathematically speaking; there is a relation; any set of number pairs is a relation._

12. Give a description of the typical player on the team using the information you have obtained about his height, weight, and grades. _He is 181-182 centimeters tall, weighs about 86-87 kilograms, and earns B grades._

2

Dealing with Uncertainty

How often are you really sure about anything? Don't you often find yourself thinking, *"Probably* it will rain tomorrow," or "There is a *chance* that our team will win the game," or *"More likely than not* Johnny will be absent again tomorrow"? Children, too, think this way, and many of them even use these same words in their own conversations. The activities in this unit are designed to help students understand these expressions and use them in a more meaningful way.

If we were to deal fully with the concepts associated with such expressions we would quickly find ourselves in a rather deep study of probability theory. We certainly do not propose such a study here; instead, we shall deal with these ideas on an informal and intuitive level. Our goal is to give the students a taste of these ideas so that they will get something of the flavor of uncertainty and learn how to handle it in an elementary way.

Just as it is easier to think clearly if you first organize your thoughts, many problems in mathematics are easier if you order the ideas presented before you attempt to solve the problem. Therefore, in this unit students are given practice in organizing and in collecting and analyzing fairly simple data. They begin by working with number patterns found in everyday situations and by manipulating objects like coins and dice. Toward the end of the unit they have a chance to make predictions based on data they have collected, organized, and analyzed on their own. It is hoped that the students will find the experiences interesting and challenging.

51

Since the experiences in this unit involve an element of chance, a complete set of answers cannot be given on the reduced worksheets inserted in the text. However, note that at the end of the unit there is a section called Probability Theory: Supplementary Information. It contains a statement about the mathematical concepts involved in each experience.

OVERVIEW

Each of the five experiences in this unit includes a detailed section called "Teacher Strategy," which is one teacher's first-person account of a procedure that proved effective with his or her class.

The following brief summary will help you decide whether these experiences meet the needs of your own class and will also give you some indication of the amount of preparation involved.

Sample student worksheets are provided in the Teaching Package. It is assumed that duplicating facilities are available to you so you can provide class lots of these worksheets.

Experience 1: Recording Last Digits and Sums of Last Two Digits

Real-life situations are devised in the classroom, and attention is focused on both events that happen often and events that happen rarely. Students then work in pairs on the worksheets. They are encouraged to make predictions, based on their own recorded observations, about which events are common and which are rare.

Materials needed: telephone directory, worksheets.

Experience 2: Tossing Coins

Students experiment with flipping one and two coins at a time, first working alone, then in pairs. They record the number of times "heads" turns up and the number of times "tails" turns up. They then make predictions based on these observations and gather additional data to test their predictions.

Materials needed: coins, worksheets.

Experience 3: Rolling Dice

Students roll one die many times to determine whether any one face turns up more often than others. They then roll a pair of dice to determine which sums are most likely to turn up in a cast of dice.

Materials needed: dice, worksheets.

Experience 4: Drawing Beads

Students try to predict the combinations of three colors of beads they are most likely to draw from a can. Then they try to determine the color of both the greatest number of beads and the least number of beads in the can.

Materials needed: large can, small cans, three different colors of beads, worksheets.

Experience 5: Tossing Objects

Students toss paper cups and thumbtacks to determine the position in which they are most likely to fall.

Materials needed: paper or plastic cups, thumbtacks, worksheets.

EXPERIENCE 1
Recording Last Digits and Sums of Last Two Digits

OBJECTIVE

The students should be able to show that certain sums of two digits occur more frequently than others. They should also be able to give some explanation of why this happens.

MATERIALS

Telephone directory
3 worksheets for each student

TEACHER STRATEGY

I like to begin this experience by calling one student to the front of the room. I ask that student to open the phone directory anywhere in the white pages and to point to any one of the phone numbers on the page, then copy this number on the chalkboard.

I ask another student to name the last two digits of this phone number. A discussion follows to make sure all the students know what I mean by phrases like "the first digit," "the last two digits," and "the sum of the last two digits."

Next I call two students to the front of the room. I give one student the phone directory and ask the second student to be the recorder. The first student opens the directory to a white page and picks a phone number on that page. He then reads the last digit of each of the next fourteen numbers. As he reads each digit the recorder writes it on the board.

Referring to the list on the board, I ask, "What is the least digit we found? What is the greatest digit we found? If we did this again, what is the least digit we could find?" Continuing in this way, I lead up to a convenient way of recording the results of such "experiments" (see fig. 1 on the following page). The recorder transfers the results of our experiments to this frequency table on the chalkboard.

I then begin asking questions about the digit found most often, least

Digit	Tally	Total
0		
1		
2		
3		
4		
5		
6		
7		
8		
9		

Fig. 1

often, and so on. These lead up to a question like, "If I pick a phone number and ask you to guess the last digit, what would you guess?" Of course there is no "right" answer. But the discussion might reveal the interesting notions some students may have about their "lucky number" and other beliefs. I take these as they come and wait for students to acquire new ideas about chance events as they continue their experiments.

You may wish to ask students to begin the individual experiments immediately after this discussion, or you may want to wait until the next day to begin this phase of the experience. When you do introduce this phase, give each student a white page from the phone directory and a copy of the worksheet for Activity 1. Each student picks a partner and the partners take turns collecting and recording data.

Ask the data recorder to guess the sum of the last two digits of a phone number and write this guess in the space provided. The partner, with closed eyes, then places a finger on the white page. Next, with eyes open, the data collector gives the actual sum for the last two digits of the number closest to his or her finger and of the nine numbers following it. The recorder writes in each of these sums and counts up and records the number of times his or her guess matched the actual sum.

Some students may make a guess like thirty-six for the sum of the last two digits of a phone number but then find out for themselves that the greatest possible sum is eighteen $(9 + 9)$. They will probably also discover that sums like nine, ten, and eleven are more likely than numbers like two and seventeen, since many more pairs of digits add up to eleven, for example, than add up to two.

Each student takes three turns as recorder and three as collector. You will find that some students begin to be better guessers as they go from one trial to the next. Some may even be able to explain, in the discussion at the end of the experience, why a guess like nine is better than a guess like seventeen. (You can get a sum of nine in ten ways: $0 + 9$, $1 + 8$, $2 + 7$, $3 + 6$, $4 + 5$, and the reverse of each. There are only two ways to get seventeen: $8 + 9$ and $9 + 8$. Refer to the table given in the supplementary information at the end of this unit for a complete list of the possible combinations of digits and their sums.)

When each student has completed three trials, he or she tallies the result of each on the chalkboard. You could put a table like that shown in figure 2 on the board for this purpose.

Number of Sums That Matched My Guess	Tally
0	
1	
2	
3	
4	
5	
6	
7	
8	
9	
10	

Fig. 2

I don't discuss the tally of actual sums that matched the guessed sum. I have found that students talk about it anyway, and after thinking things over they often say, "We could do better if you gave us another chance."

I either have the second worksheet ready for students who finish the first one quickly or I plan some kind of activity to take up slack time.

You may wish to have students do the worksheets for Activities 2 and 3 in class, or you may want to have them complete parts of them at home. I like to spread this experience over at least two days, and with slow learners I spend three days on it. Sometimes I use the first worksheet several times. I give the directions orally so I can ask students to guess the sum of the last three digits and, thus, extend the possible sum to numbers from zero to twenty-seven. Different classes react differently, but I always try to get across the idea that there can be some science in guessing. With some classes this requires more than one use of the first worksheet. [*Continued on page 60*]

RECORDING DIGITS AND SUMS OF   UNIT 2, EXPERIENCE 1
LAST TWO DIGITS

Activity 1   Name _____

Follow the directions your teacher gives you.

TRIAL 1

My guess _____

Actual sums _____ _____ _____ _____ _____

_____ _____ _____ _____ _____

My guess matched the actual sum _____ times.

TRIAL 2

My guess _____

Actual sums _____ _____ _____ _____ _____

_____ _____ _____ _____ _____

My guess matched the actual sum _____ times.

TRIAL 3

My guess _____

Actual sums _____ _____ _____ _____ _____

_____ _____ _____ _____ _____

My guess matched the actual sum _____ times.

RECORDING DIGITS AND SUMS OF
LAST TWO DIGITS

UNIT 2, EXPERIENCE 1

Activity 2   Name _____

Use your page from the telephone directory to answer the questions.

1. Put your finger on the page without looking. Move your finger to the nearest phone
   number. What are the last 2 digits? _____ and _____.

2. The sum of these digits is _____.

3. Use the 30 phone numbers following this first one. On the chart below, make a tally mark
   to show each sum you get by adding the last 2 digits of each phone number. When you
   finish, add up your tally marks and record the totals in the last column.

Sum of Last Two Digits	Tally	Total
0		
1		
2		
3		
4		
5		
6		
7		
8		
9		
10		
11		
12		
13		
14		
15		
16		
17		
18		

4. Why does the table include 0? _Because_ $0 + 0 = 0.$

5. Why is 18 the largest sum included in the table? _The largest digit
   is 9, and 9 + 9 = 18._

6. Write six pairs of digits which have the sum 5. (Count 4, 1 and 1, 4 as two different
   pairs.) $(5, 0) (0, 5) (4, 1) (1, 4) (3, 2) (2, 3)$

RECORDING DIGITS AND SUMS OF   UNIT 2, EXPERIENCE 1
LAST TWO DIGITS

Activity 3   Name _____

1. Record the license-plate numbers of 10 cars. Disregard letters—write down only the
   numerals.

   _____ _____ _____ _____ _____

   _____ _____ _____ _____ _____

   Now tally the last digits and the sums of the last 2 digits.

Last Digit	Tally of Times You Got This Digit
0	
1	
2	
3	
4	
5	
6	
7	
8	
9	

Sum of Last Two Digits	Tally of Times You Got This Sum
0	
1	
2	
3	
4	
5	
6	
7	
8	
9	
10	
11	
12	
13	
14	
15	
16	
17	
18	

2. Is there a best guess for the last digit of a license-plate number?

                  _____ yes   __X__ no

3. Is there a best guess for the sum of the last 2 digits of that number?

                  __X__ yes   _____ no

4. Which sum do you think would occur most often if you looked at thousands of license
   plates? __9__

When all students have completed the three worksheets, I lead a discussion centered around the question "What did you learn?"

EVALUATION

I do not expect all students to be able to explain why some sums are more likely than others. But I do expect many students to notice the difference between guessing the last digit and guessing the sum of the last two (or three) digits. Some say, for example, "Even though more sums are possible, I'd rather guess sums. You just can't tell anything about the last digit. But I have a good system for guessing sums." If this sort of comment occurs, Experience 1 has gone well. Students have learned from their own experiments and have picked up ideas from one another. They are ready to go on to Experience 2.

EXPERIENCE 2
Tossing Coins

OBJECTIVE

The student should be able to determine whether one side of a coin is more likely to turn up than another when the coin is flipped and allowed to fall to the table.

MATERIALS

2 coins for each student (or any object that can be easily flipped and has a uniform mass)

2 worksheets for each student

TEACHER STRATEGY

You might introduce this experience by flipping a coin. Ask your students, "How many of you think it came up heads? How many think it came up tails?" Flip the coin a few more times. Then tell them that you are going to let each of them use a coin to perform an experiment in which they are to try to determine which side is more likely to turn up. Some students may already think they can predict pretty well.

Notice that the worksheet for Activity 1 asks the student to make a prediction after only twenty flips. It is hoped that the student will alter the prediction after the next thirty flips. There is a chance that students will feel tied quite closely to their exact data—that is, think that twenty-four tosses of heads and twenty-six tosses of tails definitely mean that tails is the more likely result, rather than that both results are equally likely. Don't worry about it. These concepts take a long time to develop properly. At this time our concern is that the students are able to collect their data and, on the basis of those data, come to some conclusions.

As the students complete the first worksheet, I like to distribute the worksheet for Activity 2 and have them work on it in pairs. The members of each pair take turns at tossing coins and tallying the results on their own worksheets.

After each student has had a chance to perform at least the first fifty trials (some probably will have done a hundred), I think it is a good idea to take a look at the data collected by the class as a whole. I first make on the board three columns headed H, T, and HT and ask how many students feel that one result is more likely than the others and how many feel that they are equally likely. I sometimes also collect the individual results and put the actual tabulation on the board.

I follow up with questions such as, "How many think that if I throw heads this time, I will throw tails the next time? What if I throw tails three times in a row—would the next throw *have* to be heads?"

EVALUATION

Every student should meet with some success in this experience. If they toss the coin, record their data, and come to some conclusions based on these data, they will have done all that you can expect at this point. Some students will probably see that when they collect more data the number of heads and the number of tails tend to equalize, and hence they will see that in the long run one is no more likely to happen than the other (unless the coin is weighted). Don't be concerned if not all grasp this idea. In the experiences that are to come, students should see how more and more data usually help in making better predictions.

Students should not be expected to give good answers to Questions 2, 4, and 5 in Activity 2.

TOSSING COINS   UNIT 2, EXPERIENCE 2

Activity 1   Name _____

What happens when you flip a coin 20 times? Does the same side always land up? Try it and see. Flip your coin. Notice that one side is "heads" and the other is "tails." If heads turns up, put a tally mark in the H row below. If tails turns up, put a tally mark in the T row. Make 20 tosses, then fill in the totals.

Side	Tally	Total
H		
T		

Which side came up more often? _____ Which side do you think will turn up more often if you make 30 more flips? _____

Now flip the coin 30 more times and record your results below.

Side	Tally	Total
H		
T		

Which side came up more often this time? _____ Look at the totals from both sets of flips. If you add all the heads together and all the tails together, which turns up more often? _____

Now exchange coins with a neighbor or your teacher and flip the new coin 30 times. Record your results below.

Side	Tally	Total
H		
T		

Which side came up more often this time? _____

If you have time, make another chart and try someone else's coin. In looking over all your results, does it appear that one side comes up more often than the other? _____

TOSSING COINS   UNIT 2, EXPERIENCE 2

Activity 2   Name _____

   What happens when you flip 2 coins at the same time? Try it and see. As you complete
each flip of the 2 coins, tally your result below. Do it 40 times.

The Coins Fall	Tally	Total
H,H		
H,T or T,H		
T,T		

1. Which way did the coins fall most often? _____

2. Why do you think they fell this way most often? _____
    _____

3. Was there much difference in the number of times you got H,H and the number of

    times you got T,T? _____

4. How do you explain this? _____
    _____

5. Enter below the results you might expect to get if you tossed these 2 coins 4 000 times.

The Coins Fall	Total
H,H	1000
H,T or T,H	2000
T,T	1000

– 39 –

EXPERIENCE 3
Rolling Dice

OBJECTIVE

By throwing a die at least a hundred times, the student should be able to determine whether one number tends to turn up more often than any other number. By rolling a pair of dice one hundred times he should be able to decide which sums are more likely to be rolled and which sums are less likely to be rolled.

MATERIALS

1 pair of dice for each student
5 worksheets for each student

TEACHER STRATEGY

Have you ever felt that when you threw a die you were more likely to get one number than another? Most adults feel that way even though they know it isn't so. Before students begin the activities for this experience, I ask them if they feel they have a lucky number. Then I let them throw a die to see if their findings support their hunches.

The first activity requires students to make a prediction after only a few trials. They are then given an opportunity to change their prediction when they have more data. Some students should begin to see that a better decision about an uncertain situation can be made once more information has been collected and studied, although this is not a primary objective of the experience. However, to help the class reach this secondary objective, some time should be spent looking at the statistics the students have compiled.

I begin by taking an overall view, finding out how many students think certain numbers will come up more often. Then, on the board, I tabulate the data from the class as a whole (you may wish to use only a sample of six or seven students). I list the information in columns (see fig. 3, p. 65) and let the students find the totals—this gives them some good practice in column addition. Those students who think a particular number will come up more often are in charge of the column for that number, and the others check

Name	⚀	⚁	⚂	⚃	⚄	⚅
Pupil A . . . Pupil X						
Totals						

Fig. 3

their work. By using statistics gathered from the whole class, students can more easily see that all the results are about the same and that a difference of ten in 1000 tosses is not really very much.

Next I have my students work individually on Activity 1. When most of them are finished, we discuss the results. I then give each student two dice and ask some of them to report the sum of the numbers on the two faces they get in a cast of dice. After a short discussion I hand out the worksheets for Activities 2 and 3 and let students work individually.

The main objective of Activities 2 and 3 is for students to find through experimentation that certain sums do occur more often and others less often. Some students may decide that certain sums are equally likely (5 and 9, for example). They may even order the possible sums from the most likely to the least likely (7, 6 and 8, 5 and 9, 4 and 10, 3 and 11, 2 and 12). But do not expect most students to get this far on their own.

To enable students to get these ideas from this experience, I put a class chart listing all the possible sums (fig. 4) on the chalkboard. Each student contributes his results from Activities 2 and 3. With such a large amount of data the relative likelihoods of the different sums begin to show up more clearly (and students also get practice in adding large sums).

[Continued on page 71]

Name	2	3	4	5	6	7	8	9	10	11	12
Pupil A . . . Pupil X											
Totals											

Fig. 4

ROLLING DICE   UNIT 2, EXPERIENCE 3

Activity 1   Name _____

Throw the die. What number faces up? _____ Throw it again. What number came up this time? _____ What other numbers could come up? _____

EXPERIMENT 1: Try to find out whether one number turns up on your die more often than any other number. Do you think one will? _____ If so, which one do you expect to turn up most often? _____

The table lists all the ways a die can turn up. Every time you throw the die, tally the number that turns up in the appropriate row. Keep throwing the die until one number has 10 tally marks after it. Then find your total for each number.

Face of Die	Tally	Total
⚀		
⚁		
⚂		
⚃		
⚄		
⚅		

EXPERIMENT 2: Let's do the same thing again. Toss the die until one number comes up 10 times. Do you think it will be the same number as before? _____

Face of Die	Tally	Total
⚀		
⚁		
⚂		
⚃		
⚄		
⚅		

[Continued]

ROLLING DICE   UNIT 2, EXPERIENCE 3

Activity 1—*Continued*   Name _____

EXPERIMENT 3: Exchange dice with your neighbor or your teacher and see what results you get with the new die. Do this experiment as you did the others, tossing the die until one number has come up 10 times.

Face of Die	Tally	Total
⚀		
⚁		
⚂		
⚃		
⚄		
⚅		

Now take the information from each of your 3 charts and write the totals in the chart below. Then get the grand totals by adding the 3 totals for each face.

Face of Die	Total #1	Total #2	Total #3	Grand Total (#1 + #2 + #3)
⚀				
⚁				
⚂				
⚃				
⚄				
⚅				

Use the table you have just filled in to answer these questions.

1. Did each of the 6 numbers come up at least once in each experiment? _____

2. Look at the grand totals. Did any number turn up twice as often as another one? _____

3. If you were to throw your die once more, would you know what number to expect to turn up? _____

– 41 –

ROLLING DICE

Activity 2   Name _____

Shake the dice and roll them. What sum do you get on the two dice? _____ Shake them again and roll them. What sum did you get this time? _____ Roll the dice 100 times. Record your sums in the box below. When the same sum comes up more than once, use tally marks to show the number of times it comes up. Keep track of your throws by marking off, in the block on the right, each number as you throw. Put a line through one way for the first 50 throws and the other way for the second 50 throws. For example, on toss 1 do this: ╱; on toss 51 do this: ╳.

Sums on Dice			Number of Throws									
			1	2	3	4	5	6	7	8	9	10
			11	12	13	14	15	16	17	18	19	20
			21	22	23	24	25	26	27	28	29	30
			31	32	33	34	35	36	37	38	39	40
			41	42	43	44	45	46	47	48	49	50

In the chart below, arrange your sums from smallest to largest. After each sum put the total number of times it came up in your experiment. Did you miss any possible sums in your first 100 tosses? If so, put them in their proper place in the chart and record the number of times as 0.

Sums	Times Found in 100 Tosses	Tally	New Total
Smallest:			
Largest:			

[Continued]

ROLLING DICE   UNIT 2, EXPERIENCE 3

Activity 2—*Continued*   Name _____

Look back at your chart for this activity. What sum(s) did you get most often?
_____ How many times did you get it (them)? _____
What sum(s) did you get least often or not at all? _____

Now let's try something a little different. Try to throw sums of five:

or

Keep throwing the dice until you have thrown a sum of five 10 times. In the tally column
in the chart keep a record of all the sums that turn up while you are trying to throw 10 sums
of five.

Now look at the tally column. Which sum has the most tally marks after it? _____
Which sum has the fewest tally marks after it? _____ What sum do you think is your
"lucky sum"? _____ How many times do you think you'd have to throw to get it?
_____ Try. How many throws did it take? _____ Do you think it would take that
long to get it again? _____ Try it and see! How many throws did it take? _____

ROLLING DICE   UNIT 2, EXPERIENCE 3

Activity 3   Name _____

As you have seen, some sums seem to occur more often than others. Study your table from Activity 2. Use the first column of the table below to arrange the sums in order from those you found most often to those you found least often. Then throw the dice another 50 times and tally your results in the middle column. (Do more throws if you have time.) Then find your total for each sum.

Sum on Dice	Tally	Total
Most often:		
Least often:		

Do your sums seem to be in the right order? _____ What changes would you make in the order? _____

Do you think that if you threw your dice a few more times, your results might change? _____

Look at the final totals in your tables for Activities 2 and 3 and then answer these questions.

1. Which sum occurred most often? _____
2. Which sum occurred least often? _____
3. What sums seemed to come up about the same number of times? _____
4. The next time you roll a pair of dice, are you more likely to get a sum of 3 or a sum of 8? _____ Why? _____

_____

Although the numbers appearing in a cast of dice do come up randomly, it is hoped that through this experience the students will see that even chance favors certain results. A little experimenting and data collecting can help them to find out how the cards are stacked. Knowing this can be very helpful in making decisions.

EVALUATION

The discussion following Activity 1 will tell you whether your students have learned that any one face is as likely to turn up as another. Their worksheets for Activities 2 and 3 will tell you whether they know which sums of two faces are more likely to occur. Most students should be able to answer correctly the final question for Activity 3.

EXPERIENCE 4
Drawing Beads

OBJECTIVE

The student should be able to determine the possible color combinations he can get by drawing three beads at random from a can containing thirty beads (ten of each of three colors). Also, by drawing beads from a can containing an unknown number of beads of each color, the student should be able to make an intelligent guess about the color of the largest number of beads in the can.

MATERIALS

1 large can
1 can (such as a frozen-orange-juice can) for each student
30 beads, 10 of each of 3 different colors, for each student (marbles may be substituted)
43 red
58 blue } beads or marbles for each group of 8 students (other colors may be substituted)
43 green
2-page worksheet for each student
2-page worksheet for each pair of students

TEACHER STRATEGY

I begin this experience by placing a large can filled with beads on my desk. I tell the class the three colors of the beads in it, and then I ask one student to come up and draw three beads without looking in the can. While the student still holds the beads, I ask the class to guess what colors have been drawn. Then we look at the beads and I write the color combination on the board. I then let two or three other students draw and let the rest of the class guess again, listing on the board all the combinations drawn or suggested.

A few examples of this nature should help students see what they are to do in this experience, and they can now begin working individually on Activity 1. I give each student the two-page worksheet and a small can filled with thirty beads, ten of each of three colors.

It may take some students a hundred trials to find all ten possible combinations. You should move around the room to see how they are progressing. If a student does find all the combinations fairly quickly, suggest that he or she try to find out which combinations are more likely and which are less likely. (See ''Probability Theory'' at the end of this unit for a list of combinations and some information on their expected frequencies.)

Have students check their own work by asking one student to list on the board all the combinations he or she found. Let the others add to this list. You need not explain why certain combinations occur more often than others. If the class sees that certain ones were drawn more often than others, leave it at that.

Students are to work in pairs in Activity 2. To each pair I distribute the two-page worksheet and a can of beads. Each can has thirty-six beads of three different colors, and I pass out the four different mixtures shown in figure 5. You should label the cans so that you know the contents but students do not.

I tell the students that there is an unequal number of beads of each of

[Continued on page 77]

Can	Red Beads	Blue Beads	Green Beads
A	16	12	8
B	6	18	12
C	12	4	20
D	9	24	3

Fig. 5

DRAWING BEADS   UNIT 2, EXPERIENCE 4

Activity 1   Name _____

Draw 3 beads out of your can. What colors are they? _____
Put them back in, stir the beads around with your finger, and draw 3 more. What colors
are they? _____ Let's use letters to stand for the colors.
If you get 2 reds and 1 green, write RRG. If you get a red, a blue, and a green, write RBG.
Continue drawing 3 beads at a time without looking into the can. Record below the colors
you get each time. Then put the beads back into the can, stir them up, and draw 3 more.
Make 20 draws.

1. _____	6. _____	11. _____	16. _____
2. _____	7. _____	12. _____	17. _____
3. _____	8. _____	13. _____	18. _____
4. _____	9. _____	14. _____	19. _____
5. _____	10. _____	15. _____	20. _____

Now make a list of all the *different* color combinations you have gotten so far. Also write
down the number of times you got each combination.

Color Combination	Tally of Times You Got the Combination	Total

Now make another 20 draws using the same method and record your findings below.

21. _____	26. _____	31. _____	36. _____
22. _____	27. _____	32. _____	37. _____
23. _____	28. _____	33. _____	38. _____
24. _____	29. _____	34. _____	39. _____
25. _____	30. _____	35. _____	40. _____

[Continued]

– 45 –

DRAWING BEADS   UNIT 2, EXPERIENCE 4

Activity 1—*Continued*   Name _____

Did any combinations come up that you didn't find in your first 20 drawings? _____
If so, what are they? _____ Add these to the chart on the first page of this
activity. Also put in tally marks, so that you can see how many times you got each
combination.

Make 20 more draws and record them below.

41. _____   46. _____   51. _____   56. _____
42. _____   47. _____   52. _____   57. _____
43. _____   48. _____   53. _____   58. _____
44. _____   49. _____   54. _____   59. _____
45. _____   50. _____   55. _____   60. _____

Did you get any new combinations this time? _____ If so, what are they? _____
_____ Add them to your chart and place tally marks after each one for
as many times as you got it in draws 41–60.

Use your chart to answer these questions.

1. What color combination(s) did you draw most often? _____

2. What color combination(s) did you draw least often? _____

Answer these questions after you compare your findings with those of your classmates.

1. Did you draw all the possible combinations? _____

2. If not, which ones are you missing? _____

3. If you were to draw 3 beads at random from the can, what color(s) would you expect

   them to be? _____

– 46 –

DRAWING BEADS UNIT 2, EXPERIENCE 4

Activity 2 Name _____

Name _____

The can contains 36 beads. There is a different number of beads for each color. For example, there could be 18 red, 10 blue, and 8 green beads. Take turns making trial drawings and try to decide the color of the greatest number of beads and the color of the least number of beads. (No fair counting first!)

Draw 3 beads. What colors are they? _____ Put them back in. Do you think that one drawing is enough to let you decide the color of the greatest number of beads? _____

Make 10 draws of 3 beads each, returning the beads before each new drawing. List the color combinations that you get.

1. _____ 3. _____ 5. _____ 7. _____ 9. _____
2. _____ 4. _____ 6. _____ 8. _____ 10. _____

What color do you think most of the beads are? _____

CONCLUSION 1: There are fewer _____ beads and more _____
beads than any other color. (color) (color)

Make 10 more draws of 3 beads at a time and see if these results make you want to change Conclusion 1.

1. _____ 3. _____ 5. _____ 7. _____ 9. _____
2. _____ 4. _____ 6. _____ 8. _____ 10. _____

[Continued]

DRAWING BEADS   UNIT 2, EXPERIENCE 4

Activity 2—*Continued*   Name _____

                                        Name _____

Look at your results so far. Then complete Conclusion 2.

CONCLUSION 2: We think that there are fewer _____ beads and more
_____ beads than any other color.

Now make one more check.

1. _____   3. _____   5. _____   7. _____   9. _____

2. _____   4. _____   6. _____   8. _____   10. _____

CONCLUSION 3: Looking at all 30 drawings, we think there are fewer _____
beads and more _____ beads than any other color.

If Conclusions 1, 2, and 3 agree, or if 2 and 3 agree, record your Final Decision below and
ask your teacher for another can to decode. If Conclusions 2 and 3 do not agree, make at
least 20 more drawings before making your final decision.

1. _____   5. _____   9. _____   13. _____   17. _____

2. _____   6. _____   10. _____   14. _____   18. _____

3. _____   7. _____   11. _____   15. _____   19. _____

4. _____   8. _____   12. _____   16. _____   20. _____

FINAL DECISION: The color of the greatest number of beads is _____.

                    The color of the fewest number of beads is _____.

(We used the can labeled _____ to make our decision.)

three colors in their can and that by drawing enough samples of three beads they are to determine the color of the largest number of beads and the color of the least number of beads. Partners take turns drawing beads and recording results on the worksheet. When they complete the activity, I have them make a count of the various colors of beads in the can to check their final decision. You may wish to have the first pairs to finish exchange their cans of beads and try to determine the color of the largest number of beads in the new can by the same sampling method.

EVALUATION

Most students should be able to determine the ten color combinations they can get in Activity 1. The class discussion and the students' work in Activity 2 should help you determine whether most of your students can make accurate predictions from the samples drawn.

EXPERIENCE 5
Tossing Objects

OBJECTIVE

By collecting their own experimental evidence, the students should be able to determine which way physical objects such as paper cups and thumbtacks are most likely to land when tossed.

MATERIALS

1 paper or styrofoam cup for each student
1 thumbtack for each student
3 worksheets for each student

───── TEACHER STRATEGY ─────

I like to begin this experience by asking my students which way they would expect a cup with a flat bottom to land if they were to toss it 30 centimeters into the air and let it land on their desks. I record on the board the number of students who predict each of the three possible outcomes. Then I pass out the two-page worksheet for Activity 1 and a paper cup to each student. I ask the students to read the directions carefully and to work individually while completing the worksheet.

Which way do you think it will land most often? If you're not sure, toss a cup a few times until you decide on your answer—there is no mathematical answer to this question. I don't know what answer your students will come up with. It will depend upon the kind of cup used.

When most of the students have made a final prediction, put the class data on the board and come to a class decision. If a student disagrees quite strongly with the class results, encourage him or her to make some more trials. (Perhaps the decision was based on too little evidence.) If the student still does not agree suggest that it may be the particular cup used—perhaps it is a little different from the others. Let the student try another cup and see what results are obtained.

Then I pass out the thumbtacks and worksheets for Activity 2 and direct the class to go to work independently. It is important that each student collect a sufficient amount of data (at least a hundred tosses) before making a decision. Also, a student's decision should follow from his or her own data—this is much more important than reaching a decision that agrees with the results of the rest of the class.

When all students have completed Activity 2, I discuss with them factors that could bring differing results. Some of these factors are the particular tack used, the height from which it is thrown, the force of the throw, and so on. Perhaps they will have other suggestions. Ask them what they could do to make their experiments more scientific, as if they were collecting scientific data and making predictions based on the data. Ask some students to come to the front of the class to demonstrate the collection of data under better-controlled conditions. For example, one of them might always drop the tack from a height of 30 centimeters.

EVALUATION ───────────────────────────

There are two things to look for in evaluating the students' work in both activities. First, did they make a sufficient number of trials before making a decision? Second, do their conclusions follow from the data collected?

TOSSING OBJECTS   UNIT 2, EXPERIENCE 5

Activity 1   Name _____

Which way do you think a paper or plastic cup will land if you just toss it into the air and let it fall? Put an X under the position you think it will land in most often.

_____   _____   _____

Put a 0 under the way you think it will land least often.

Now toss your cup a few times and see what happens. Make a tally mark in one of the columns to record the way it lands each time. Toss the cup 20 times.

Bottom	Top	Side

Would you like to make a prediction now? If so, circle your prediction.

Most Likely   **Least Likely**

PREDICTION 1:

Make another 20 trials and record your results here.

Bottom	Top	Side

[Continued]

TOSSING OBJECTS   UNIT 2, EXPERIENCE 5

Activity 1—*Continued*   Name _____

Study the information from both experiments and then make your second prediction.

Most Likely   **Least Likely**

PREDICTION 2:

Make some more tosses. Make at least 20, but more if you have time.

Bottom	Top	Side

FINAL PREDICTION: I have made _____ tosses of the cup.

On the basis of these tosses, I have decided the following.

Most Likely   **Least Likely**

TOSSING OBJECTS   UNIT 2, EXPERIENCE 5

Activity 2   Name _____

Have you ever tossed a thumbtack in the air and had it land point down, sticking into the floor? _____ It doesn't happen very often, does it? How do you think a thumbtack will usually land, like this ⊿ or like this ⊽ ? Circle one.

EXPERIMENT 1: Toss the tack 25 times. Show the way that it lands each time by placing a tally mark in the correct column of row 1.

Experiment	⊿ Lands Up	⊘ Lands on Side	⊽ Lands Down
1			
2			
3			

Which way did it land most often? _____

EXPERIMENT 2: Make another 25 tosses and record them in row 2. Then look at the results of your first 50 tosses. Which way has the tack landed most often now? _____ Did it come up one way twice as often as any other way?_____ If so, which way? _____ Do you think you can predict the way that this tack will usually land? _____ If so, which way? _____

EXPERIMENT 3: If you didn't make a prediction after Experiment 2, do another 50 tosses and record the results in row 3.
If you decided to make a prediction after Experiment 2, check it by doing the 50 tosses.

FINAL CONCLUSION: I have come to the following conclusion based on my tosses.
The most likely way for a thumbtack to land is:

⊿   ⊘   ⊽

My data indicate that this thumbtack is no more likely to
land one way than another. _____
Would a different thumbtack make any difference? _____

- 51 -

PROBABILITY THEORY
Supplementary Information

The data in this supplementary section will help you evaluate the results your students obtain in the activities for this unit. We discuss theoretical distributions of certain digits and sums of digits in telephone numbers; of heads and tails when coins are tossed; of the faces on a pair of dice; and of color combinations of beads drawn from a can containing three colors of beads. To convince yourself that the results in practice may vary considerably from the theoretical results, you are urged to perform these experiments in advance and record the outcomes.

Experience 1. Note first that we have not tried to make a careful distinction between a number and a numeral, which is a name or a symbol for a number. Since digits are numerals, not numbers, it is incorrect to speak of "sums of digits" as we did in this experience when referring to the last two digits of the numerals in a telephone book. However, incorrect usages have become so firmly established that we hesitate to insist on the strictly correct language. Wouldn't it surprise you to hear an operator say, "Numeral please"?

If the telephone numbers (represented by numerals) are fairly randomly distributed, we can expect each digit to occur as the last digit in the numerals on any page of the directory one-tenth of the time. But the following distribution chart (fig. 6, p. 83) shows rather vividly why the eighteen possible sums of numbers represented by the last two digits in telephone numbers have different frequencies. The numerals connected by the lines in the table show that we would expect a sum of twelve 7 times in 100 sums, whereas we would expect a sum of sixteen 3 times in 100 sums of two single-digit numerals. Not only does this chart show the relative frequency of any sum, but also it shows the addition combinations that give that sum.

Experience 2. If the coin tossed is fair—that is, is symmetrical—we can expect as many heads as tails in a long series of tosses. Avoid the temptation to expect a higher probability of a tail on the toss following an unlikely run of, say, ten heads in a row. The coin has no memory, and if it is a fair coin the probability of a tail on the next toss is always $\frac{1}{2}$.

In tossing two coins the outcomes H,H; H,T; T,H; and T,T are equally likely, so we can expect each outcome about one-fourth of the time. Since we have grouped two outcomes as "H,T or T,H" on the tables in Activity 2, we can expect about half of the tally marks to be placed in this row.

+	0	1	2	3	4	5	6	7	8	9
0	0	1	2	3	4	5	6	7	8	9
1	1	2	3	4	5	6	7	8	9	10
2	2	3	4	5	6	7	8	9	10	11
3	3	4	5	6	7	8	9	10	11	12
4	4	5	6	7	8	9	10	11	12	13
5	5	6	7	8	9	10	11	12	13	14
6	6	7	8	9	10	11	12	13	14	15
7	7	8	9	10	11	12	13	14	15	16
8	8	9	10	11	12	13	14	15	16	17
9	9	10	11	12	13	14	15	16	17	18

Fig. 6

Experience 3. Assuming that a die is not loaded, we can expect any of its six faces to come up equally often. The probability of getting any particular sum from two to twelve can be displayed by making a chart similar to that shown in figure 6. Just omit 0, 7, 8, and 9 in the two borders and the corresponding portions of each row and column.

Experience 4. This experience involves some rather difficult but important concepts of probability and statistical inference. Although they of course will not be discussed with the students, a few of these ideas are mentioned here so you can better understand the results students are likely to get.

The probability of drawing three red beads from a can containing ten red, ten blue, and ten green beads is

$$\frac{10 \cdot 9 \cdot 8}{1 \cdot 2 \cdot 3} \div \frac{30 \cdot 29 \cdot 28}{1 \cdot 2 \cdot 3}, \quad \text{or} \quad \frac{120}{4{,}060}.$$

There are 4,060 distinct combinations of three articles chosen from thirty, and there are 120 ways to pick three red beads from the ten red beads.

Similarly, the ways in which we get the color combination RBB are

$$10 \cdot \frac{10 \cdot 9}{1 \cdot 2}, \quad \text{or} \quad 450.$$

Figure 7 shows the probability of each color combination both as a fraction and as a decimal approximation. The sum of these probabilities, with an allowance for rounding off, is one, as it should be. The chart indicates that in 1,000 drawings we could expect about 30 RRRs, 111 RRBs, 246 RBGs, and so on. Of course it could happen that one or more of these color combinations would not occur in the sixty drawings required for Activity 1.

RRR	RRB	RRG	RBB	RBG	RGG	BBB	BBG	BGG	GGG
$\frac{120}{4,060}$	$\frac{450}{4,060}$	$\frac{450}{4,060}$	$\frac{450}{4,060}$	$\frac{1,000}{4,060}$	$\frac{450}{4,060}$	$\frac{120}{4,060}$	$\frac{450}{4,060}$	$\frac{450}{4,060}$	$\frac{120}{4,060}$
.030	.111	.111	.111	.246	.111	.030	.111	.111	.030

Fig. 7

The probability of drawing the various color combinations in each of the four mixtures suggested for Activity 2 are given in figure 8.

After students list the color combination for each drawing, they will probably count the number of red, blue, and green beads in ten drawings and compare these totals in deciding the color of the largest number of beads. As the number of drawings increases, the ratio of a particular bead drawn tends to get closer to the exact ratio of that bead to the total number of beads in the can.

Experience 5. We have no theoretical results for either object. The combined results of students' experiments will give the best approximation to the various probabilities.

Mixture			Probability									
R	B	G	RRR	RRB	RRG	RBB	RBG	RGG	BBB	BBG	BGG	GGG
16	12	8	.078	.202	.135	.148	.215	.063	.031	.074	.048	.008
6	18	12	.003	.038	.025	.130	.181	.055	.114	.257	.166	.031
12	4	20	.031	.037	.185	.010	.135	.319	.001	.017	.107	.160
9	24	3	.012	.121	.015	.348	.091	.003	.284	.116	.010	.000

Fig. 8

Activity 1   Name _____

The students in Classroom 204 turned in a record of their birthdays. Here they are:

Jan. 20	Oct. 20	Feb. 14	Aug. 16	Feb. 12
Aug. 15	Nov. 12	Aug. 18	Jan. 2	Dec. 20
Apr. 12	Apr. 15	Mar. 20	Aug. 14	Jan. 15
Sept. 10	Jan. 9	Oct. 10	June 9	Oct. 14
Feb. 15	Sept. 22	June 9	Nov. 11	Jan. 20

Use this information to complete the frequency tables below. Then answer the questions.

Month	Tally	Frequency	Day	Tally	Frequency	Day	Tally	Frequency	Day	Tally	Frequency
Jan.			1			11			21		
Feb.			2			12			22		
Mar.			3			13			23		
Apr.			4			14			24		
May			5			15			25		
June			6			16			26		
July			7			17			27		
Aug.			8			18			28		
Sept.			9			19			29		
Oct.			10			20			30		
Nov.											
Dec.											

What month has the most births? _____ How many? _____

Do any months have no births? _____ If so, which one(s)? _____

How many months have at least 3 births? _____

How many have less than 2? _____

On what day of the month did most births occur? _____

On what days were there no births? _____

_____

_____

Activity 2   Name _____

The data below came from a survey of the number of brothers and sisters of students in a certain class. "B" means brothers and "S" means sisters, so that 2B 3S means 2 brothers and 3 sisters.

2B 3S	2B 3S	3B 1S	1B 3S
3B 2S	0B 0S	3B 3S	1B 4S
1B 0S	2B 0S	2B 2S	3B 4S
4B 2S	0B 2S	0B 0S	1B 4S
1B 1S	2B 5S	3B 1S	2B 3S
3B 1S	4B 0S	4B 1S	0B 4S
2B 2S	0B 2S	2B 1S	6B 1S
1B 3S	2B 0S	2B 2S	1B 2S

Complete the frequency tables below and then answer the questions that follow.

Number of Brothers	Tally	Frequency	Number of Sisters	Tally	Frequency	Total Number of Both	Tally	Frequency

The greatest number of brothers anyone has is _____.

The greatest number of sisters is _____.

The least number of brothers is _____.

The least number of sisters is _____.

What is the most frequent number of brothers? _____

Of sisters? _____

Of brothers and sisters? _____

Activity 1, Station A   Name _____

SKYDIVER OVER TARGET

   Stand on a chair next to the target. Drop the five squares, one at a time, onto the target from shoulder height. Do this three times.

Individual Data Sheet

Target Area	Trial 1		Trial 2		Trial 3	
	Tally	Score	Tally	Score	Tally	Score
5						
4						
3						
2						
1						
0						
Score Totals						

   To obtain the score, multiply the number of the target area by the number of squares that landed there. For example, if the frequency for Target Area 5 is 2, the score to be entered for that area would be 10.

Score for Trial 1 _____

Score for Trial 2 _____

Score for Trial 3 _____

Activity 1, Station A

Name _____

Name _____

Name _____

SKYDIVER OVER TARGET

Team Data Sheet

Summary of Scores

Name	Trial 1	Trial 2	Trial 3

Distribution of Scores for the Team

Scores	Tally	Frequency
0–4		
5–9		
10–14		
15–19		
21–25		

Activity 2, Station B   Name _____

BUTTON TOSS

Stand behind the line and toss the five buttons, one at a time, at the target. Measure the distance each button lands from the target and enter the tallies under Trial 1. Repeat for Trials 2 and 3. To obtain your score multiply the number of tallies by the points for each tally.

Individual Data Sheet

Points for Each Tally	Distance from Target in Cm	Trial 1		Trial 2		Trial 3	
		Tally	Score	Tally	Score	Tally	Score
10	0–10						
9	11–20						
8	21–30						
7	31–40						
6	41–50						
5	51–60						
4	61–70						
3	71–80						
2	81–90						
1	91–100						
0	101 or more						
	Total Scores						

Score for Trial 1 _____

Score for Trial 2 _____

Score for Trial 3 _____

Activity 2, Station B

Name _____

Name _____

Name _____

BUTTON TOSS

Team Data Sheet

Summary of Scores

Team	Trial 1	Trial 2	Trial 3
Team Member 1			
Team Member 2			
Team Member 3			

Distribution of Scores for the Team

Scores	Tally	Frequency
0–4		
5–9		
10–14		
15–19		
20–24		
25–29		
30–34		
35–39		
40–44		
45–49		
50		

Activity 3   Name _____

VOWEL COUNT

From your English classes, you are already familiar with the vowels *a, e, i, o,* and *u.* I'll bet you never expected to use them in math. Actually, all we're trying to do is find out which ones are used the most.

Take the clipping out of your "Vowel Count" envelope.

Look at the first word. Suppose it is "today." For that word you would put a tally in the *o* column and another tally in the *a* column. Do this for every word in the sentences you have. When you have finished, find the total number of tallies for each vowel and place that number in the last column. Then place your results in the proper place on the sheet for the class report.

Vowel	Tally	Frequency
a		
e		
i		
o		
u		

Activity 4   Name _____

HORSESHOE GAME

The object of this game is to move your marker from start to finish in as few moves as possible.

Place your marker at the starting line. Toss the die. Move the marker forward along the horseshoe as many spaces as the top of the die indicates, and record a tally mark in the Tally of Tosses column on your Individual Data Sheet for this activity. Continue tossing the die, moving forward the indicated number of spaces, and recording a tally mark for each toss.

You must cross the finish line on an exact count. If your die toss results in a number more than the number of spaces that remain you must keep your marker where it is, record a tally for the toss, and toss again (with a tally each time) until you get a number that you can use.

You may continue to play the game until the warning signal is given by your teacher. Then complete the frequency table on the Individual Data Sheet and record your data on the table for the class.

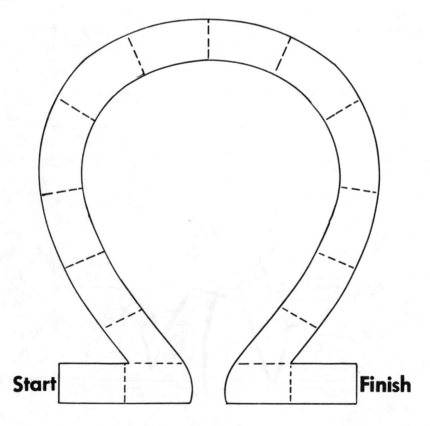

Start   Finish

Activity 4

Name _____

HORSESHOE GAME

Individual Data Sheet

Trial	Tally of Tosses	Frequency
1		
2		
3		
4		
5		
6		
7		
8		
9		
10		
11		
12		
13		
14		
15		
16		
17		
18		
19		
20		

Summary of Trials

Number of Tosses	Tally	Frequency
3		
4		
5		
6		
7		
8		
9		
10		
11		
12		
13		
14 or More		

Activity 1   Name _____

Follow the directions given in class to construct a bar graph to represent the class data from the Skydiver over Target game.

Frequency of Scores

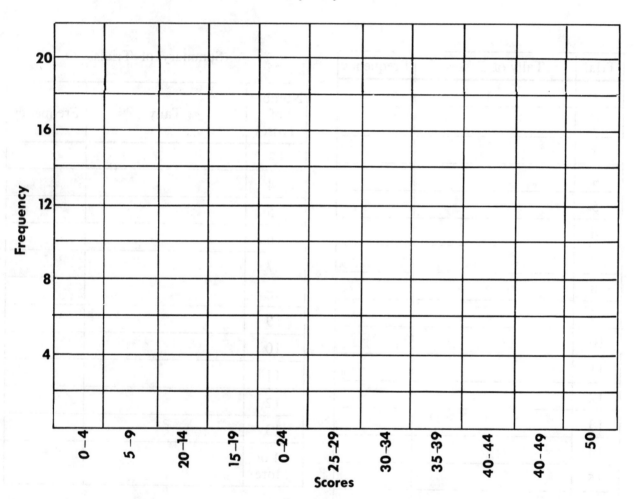

Activity 2, Station A   Name _____

LETTER COUNT

Take a paragraph out of one of the Letter Count envelopes.

Look at the first word. Suppose it is the word *when*. *When* is a 4-letter word, so you should put a tally mark after the 4 in the Tally column of the table below. Use hyphenated words as one word.

When you have tallied the number of letters for each word in your paragraph, find the total number of tallies for each number of letters and put that number in the last column.

Individual Data Sheet

Number of Letters in Word	Tally	Frequency
1		
2		
3		
4		
5		
6		
7		
8		
9 or More		

Enter these results in your group data report, then go on to the next page.

[Continued]

Activity 2—*Continued*   Name _____

Make a graph of the information in your Individual Data Sheet.

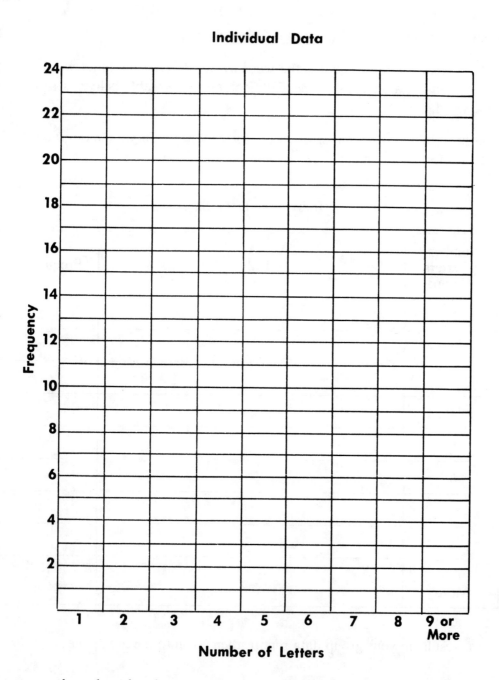

Individual Data

When your group data sheet has been completed, make a graph of the information. No form is provided.

Teacher Materials

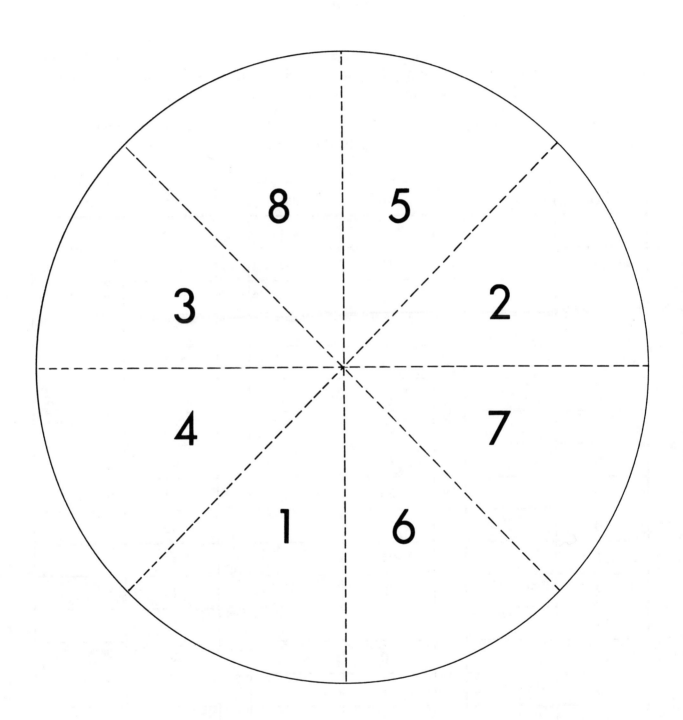

Activity 3, Station B   Name _____

SPINNER SUMS

Pick up a spinner from Station B and take it to your desk. Keep it flat on your desk and spin the pointer once. Look at the number that the pointer indicates. If a pointer stops on a line, spin it again. Then spin it again to get a second number.

Find the sum of the numbers. Write the sum after Trial 1, below. Continue spinning the pointer and finding the sums until you have filled in all the spaces in the charts below.

Trial	Sum	Trial	Sum	Trial	Sum
1		17		33	
2		18		34	
3		19		35	
4		20		36	
5		21		37	
6		22		38	
7		23		39	
8		24		40	
9		25		41	
10		26		42	
11		27		43	
12		28		44	
13		29		45	
14		30		46	
15		31		47	
16		32		48	

[*Continued*]

Activity 3—*Continued*   Name _____

Look at your data on the preceding page. What is the smallest sum you got? Put it on the first line below. What is the largest sum? Put it on the bottom line. Fill the other sums in, in order from smallest to largest. Record tally marks after them to indicate the number of times you obtained each sum.

Individual Data Sheet

	Sum	Tally	Frequency
Smallest Sum =			
Largest Sum =			

Record the above data on your Group Data Sheet.

[Continued]

Activity 3—*Continued*   Name _____

Complete this chart just as you did when you organized your own data except that this time you should use the data reported by your group on the group data report form. When you have completed the chart, use the information to construct a graph in the grid provided on the next page.

Group Data Sheet

Sum	Frequency
2	
3	
4	
5	
6	
7	
8	
9	
10	
11	
12	
13	
14	
15	
16	

[*Continued*]

Activity 3—*Continued*   Name _____

Graph of Group Data

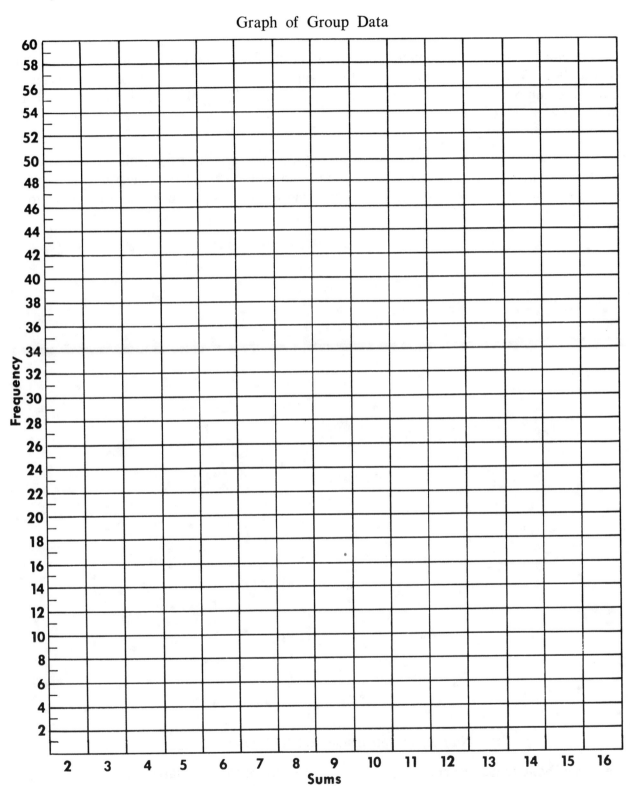

Activity 4   Name _____

DOT COUNT

   Hold your hand over the bottom part of this sheet. Close your eyes. Drop your hand, flat, on the sheet. Trace around your hand with a pencil. Stop at the base of your palm. Then count the number of dots in the area your hand covered. To make counting easier, do it by areas. For example, count the dots in each of the fingers and then divide the palm into regions and count them.

   After you have counted the dots, measure the distance from the tip of your longest finger to the base of your palm as you have traced your hand on the paper. Also measure the length of one of your shoes. Enter this information in the spaces provided below and on your group report sheet.

Number of dots inside sketch of hand _____

Length of hand to nearest cm _____ Length of shoe to nearest cm _____

Activity 5   Name _____

Study the tables and graphs you have made in the activities in this experience, then use them to help you answer the following questions:

1. How many different sums did you find, using the two spinners? _____

2. Did you get some of the sums more often than others? _____

3. What sum(s) did you get most often? _____

4. What sum(s) did you get least often? _____

5. Did you ever get a sum of 20? _____ If not, why not? _____

    _____

6. How many letters were there in the shortest word you had? _____

7. How many letters were there in the longest word you had? _____

8. Do you think that you might have had different results if you had used a different magazine? _____

9. In general most words in the paragraph are   3   4   5   6   letters in length. Circle the number(s) that apply.

10. Do you think anyone else's Dot Count will be exactly the same as yours? _____

11. Do you think your Dot Count will be high, low, or in the middle of the data collected by the entire class? _____

12. Do you think that the size of your hand makes any difference in the number of your Dot Count? _____

Teacher Materials

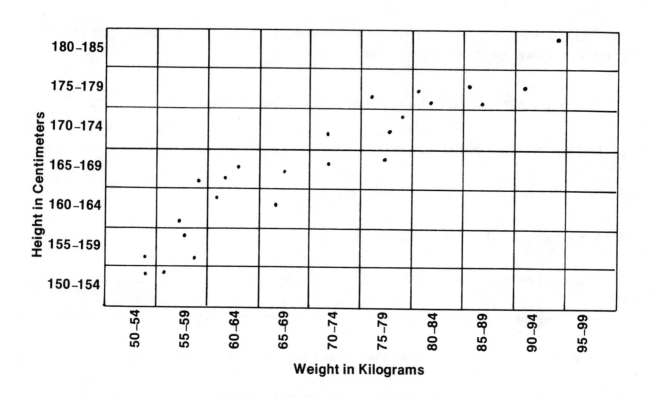

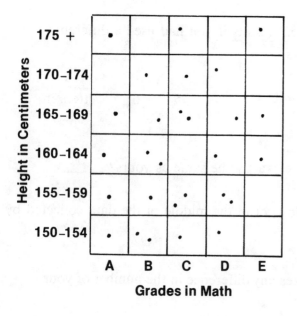

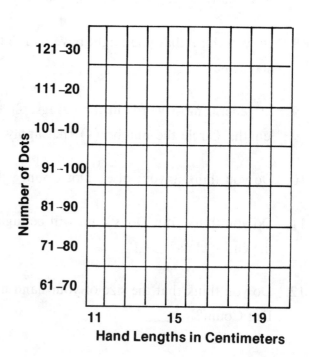

Activity 1   Name _____

BUTTON TOSS

Complete the Individual Data charts, then have one member of your group complete the Group Data Sheet on the next page.

Individual Data

Line A (0.5 m)

Distance in Cm	Tally	
	1	2
0–10		
11–20		
21–30		
31–40		
41–50		
51–60		
61–70		
71–80		
81–90		
91–100		
101 or More		

Scores _____ _____

Line B (1.0 m)

Distance in Cm	Tally	
	1	2
0–10		
11–20		
21–30		
31–40		
41–50		
51–60		
61–70		
71–80		
81–90		
91–100		
101 or More		

Scores _____ _____

Line C (1.5 m)

Distance in Cm	Tally	
	1	2
0–10		
11–20		
21–30		
31–40		
41–50		
51–60		
61–70		
71–80		
81–90		
91–100		
101 or More		

Scores _____ _____

Score Key

Distance in Cm	Points	Distance in Cm	Points	Distance in Cm	Points
0–10	10	41–50	6	81–90	2
11–20	9	51–60	5	91–100	1
21–30	8	61–70	4	101 or More	0
31–40	7	71–80	3		

[*Continued*]

Activity 1—*Continued*   Name _____

| Line D (2.0 m) | Line E (2.5 m) | Line F (3.0 m) |

Distance in Cm	Tally		Distance in Cm	Tally		Distance in Cm	Tally	
	1	2		1	2		1	2
0–10			0–10			0–10		
11–20			11–20			11–20		
21–30			21–30			21–30		
31–40			31–40			31–40		
41–50			41–50			41–50		
51–60			51–60			51–60		
61–70			61–70			61–70		
71–80			71–80			71–80		
81–90			81–90			81–90		
91–100			91–100			91–100		
101 or More			101 or More			101 or More		

Scores ____ ____   Scores ____ ____   Scores ____ ____

One of your group members should make a summary of the scores in the report below.

Group Data Sheet

Line A	Line B	Line C	Line D	Line E	Line F

[*Continued*]

Activity 1—*Continued*   Name _____

Complete the chart and the graph below.

Group Results

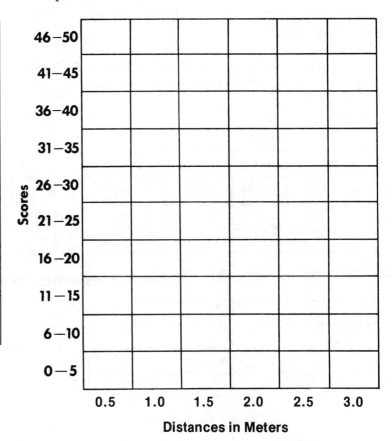

Scores	
A Line (0.5 m)	
B Line (1.0 m)	
C Line (1.5 m)	
D Line (2.0 m)	
E Line (2.5 m)	
F Line (3.0 m)	

According to your group's graph, did there appear to be any relation between the distances you stood from the target line and the scores you made? _____

If so, what relation did you notice? _____

_____

From which line were the highest scores made? _____

From which line were the lowest scores made? _____

Activity 2   Name _____

SPINNER RELATIONS

Place your spinner flat on your desk. Spin the pointer once and record the number in the first column of table 1. Spin it again and record the number in the second column of the same table. Then, in table 2, record the number for Spin 1 and the sum of the numbers on the two spins.

Continue doing this, recording the results in the two tables, until you have completely filled in all the spaces. Then go on to the next page.

Table 1

Numbers on Spinner		Numbers on Spinner	
Spin 1	Spin 2	Spin 1	Spin 2

Table 2

Spin 1 and Sums		Spin 1 and Sums	
Spin 1	1 and 2	Spin 1	1 and 2

[Continued]

Activity 2—*Continued*   Name _____

Use the pairs of numbers that you collected for table 1 of Spinner Relations to construct the graph below.

For each pair of numbers there should be a dot on the graph. For example: If the first pair of numbers was a 5 on Spin 1 and a 2 on Spin 2, you should place a dot, as shown, in the space where the fifth vertical column and the second horizontal column intersect.

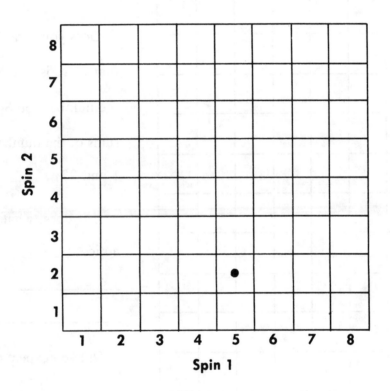

Answer the questions below, then go on to the next page.

Does your graph seem to indicate that there is any relation between the numbers on

Spin 1 and Spin 2? _____

If so, what relation do you notice? _____

_____

Did you expect to find a relation? _____

[Continued]

Activity 2—*Continued*   Name _____

Use the same procedure as before, but this time use the data from table 2 and look for a relation between the numbers on Spin 1 and the sum of the numbers on Spins 1 and 2.

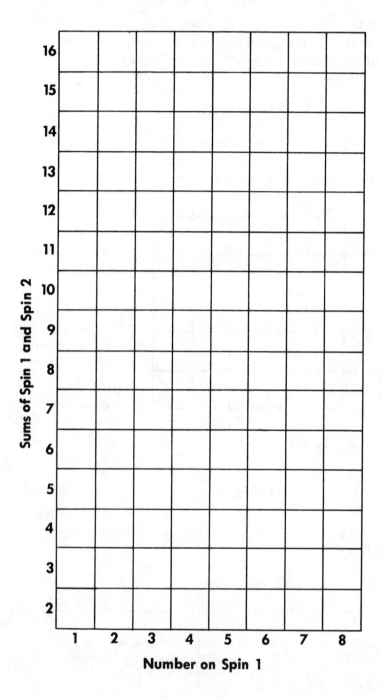

Does your graph seem to indicate that there is any relation between the numbers on Spin 1 and the sums of the numbers of Spins 1 and 2? _____

If it does, what relation do you notice? _____

_____

_____

Did you expect to find a relation? _____

Do you think that there would be a relation between the numbers on Spin 2 and the sums of the numbers of Spins 1 and 2? _____

Activity 3   Name _____

WORD LENGTHS AND NUMBER OF VOWELS

Individual Data

How do you think the number of vowels a word has compares with the number of letters it has? The longer the word the more vowels? How about the words *a* and *check?* Both words have only 1 vowel, but one has 1 letter and the other has 5. Perhaps these are just exceptions?

Take a paragraph from the envelope. Look at the first word. Put the number of letters it has in the top space of the first column and the number of vowels in the top space of the second column. Make one entry in the chart for each word in your paragraph.

Make a graph of the data in the chart on the grid provided on the next page. There should be one dot in your graph for each pair of numbers that appears in your chart.

When the graph is completed, look to see if there seems to be any relation between the number of letters in the word and the number of vowels it has. Use the space provided under the graph to describe what you find.

Letters in Word	Vowels in Word	Letters in Word	Vowels in Word

[Continued]

Activity 3—*Continued*   Name _____

WORD LENGTHS AND NUMBER OF VOWELS

Relation Graph

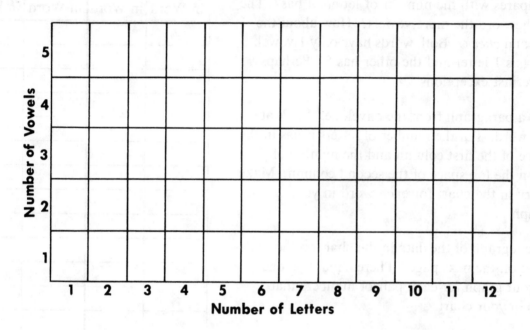

Describe the relation that seems to be shown by the points on your graph. If there seems to

be no relation, say so and explain why. _____

_____

_____

_____

Football Roster

STATE UNIVERSITY

Number	Height in Cm	Weight in Kg	Grade Average
15	184	77	A
70	183	92	B
72	185	95	B
67	181	92	C
43	180	88	C
61	181	92	B
78	183	99	A
74	180	91	B
76	183	97	C
38	177	79	B
51	188	86	C
28	183	81	A
27	178	78	D
30	175	79	B
54	175	92	B
53	182	90	C
25	179	76	B
75	182	92	A
60	179	88	D
20	182	77	A

Number	Height in Cm	Weight in Kg	Grade Average
47	183	90	B
24	181	78	C
22	175	74	B
19	181	78	B
42	178	85	A
34	186	79	D
36	178	88	D
18	181	79	B
65	185	95	C
31	178	75	B
62	173	90	C
33	180	82	A
83	188	104	A
49	180	82	B
88	190	108	B
58	179	86	C

Activity 1   Name _____

Using data from the Football Roster sheet, complete the two graphs below and a third one, at the top of the next page.

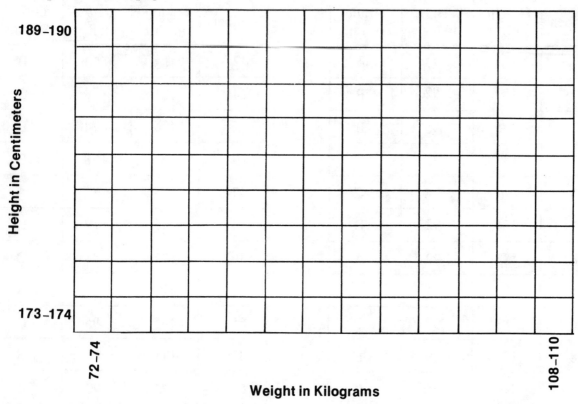

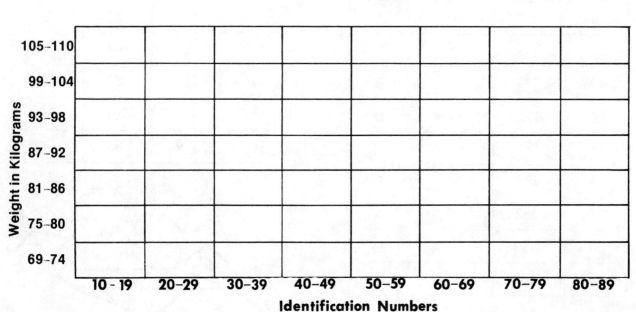

[Continued]

Activity 1—*Continued*   Name _____

Grades and Weight

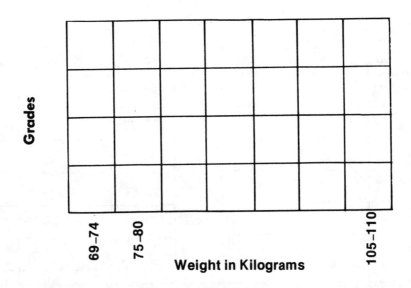

Use the grid at the right
to make any other
comparison that you
would like to consider.

Activity 1—*Continued*   Name _____

Using the data sheet again, complete the frequency tables below.

	Height	Tally	Frequency
Shortest =	173–174		
Tallest =	189–190		

Weight	Tally	Frequency
72–74		
75–77		
78–80		
81–83		
108–110		

Grade	Tally	Frequency

[*Continued*]

Activity 1—*Continued*   Name _____

Using the frequency tables you have just made, complete the graphs below.

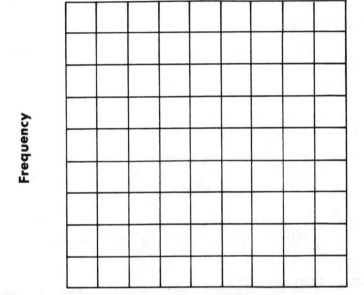

Heights in Centimeters

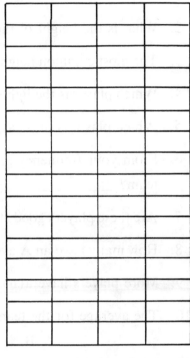

Grades

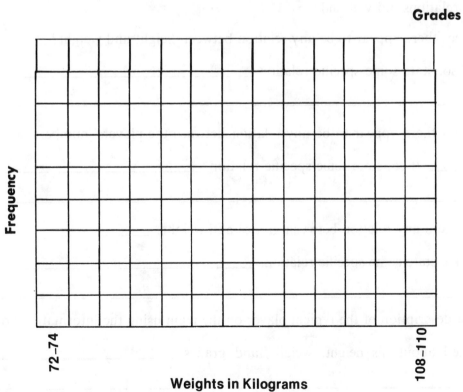

Weights in Kilograms

Activity 2   Name _____

Summary Sheet

1. What is the height of the tallest player? _____

2. What is the height of the shortest player? _____

3. The most common height is _____.

4. Which player is the lightest? _____

5. Which player is the heaviest? _____

6. From your frequency graph, what would you consider to be an average weight for the team? _____

7. Are these players good students? _____

8. How many have an A average? _____

9. More players have a(n) _____ average than any other.

10. The average for the team is in which range? Circle one.

    A–B   B–C   C–D   D–F

11. What relations did you find?

    a) Does there appear to be any relation between height and weight? _____

        If so, state some specific relation. _____

        _____

    b) Does there appear to be any relation between the players' numbers and their weight?

        _____ If so, state some specific relation. _____

        _____

    c) Is there any relation between grades and weight? _____

        If so, state some specific relation. _____

        _____

12. Give a description of the typical player on the team using the information you have

    obtained about his height, weight, and grades. _____

    _____

Activity 1

Name _____

Follow the directions your teacher gives you.

TRIAL 1

My guess _____

Actual sums _____ _____ _____ _____ _____

_____ _____ _____ _____ _____

My guess matched the actual sum _____ times.

TRIAL 2

My guess _____

Actual sums _____ _____ _____ _____ _____

_____ _____ _____ _____ _____

My guess matched the actual sum _____ times.

TRIAL 3

My guess _____

Actual sums _____ _____ _____ _____ _____

_____ _____ _____ _____ _____

My guess matched the actual sum _____ times.

Activity 2   Name _____

Use your page from the telephone directory to answer the questions.

1. Put your finger on the page without looking. Move your finger to the nearest phone

   number. What are the last 2 digits? _____ and _____.

2. The sum of these digits is _____.

3. Use the 30 phone numbers following this first one. On the chart below, make a tally mark
   to show each sum you get by adding the last 2 digits of each phone number. When you
   finish, add up your tally marks and record the totals in the last column.

Sum of Last Two Digits	Tally	Total
0		
1		
2		
3		
4		
5		
6		
7		
8		
9		
10		
11		
12		
13		
14		
15		
16		
17		
18		

4. Why does the table include 0? _____

5. Why is 18 the largest sum included in the table? _____

   _____

6. Write six pairs of digits which have the sum 5. (Count 4, 1 and 1, 4 as two different

   pairs.) _____

Activity 3   Name _____

1. Record the license-plate numbers of 10 cars. Disregard letters—write down only the numerals.

   _____ _____ _____ _____ _____

   _____ _____ _____ _____ _____

   Now tally the last digits and the sums of the last 2 digits.

Last Digit	Tally of Times You Got This Digit
0	
1	
2	
3	
4	
5	
6	
7	
8	
9	

Sum of Last Two Digits	Tally of Times You Got This Sum
0	
1	
2	
3	
4	
5	
6	
7	
8	
9	
10	
11	
12	
13	
14	
15	
16	
17	
18	

2. Is there a best guess for the last digit of a license-plate number?

   _____ yes   _____ no

3. Is there a best guess for the sum of the last 2 digits of that number?

   _____ yes   _____ no

4. Which sum do you think would occur most often if you looked at thousands of license plates? _____

Activity 1 Name _____

What happens when you flip a coin 20 times? Does the same side always land up? Try it and see. Flip your coin. Notice that one side is "heads" and the other is "tails." If heads turns up, put a tally mark in the H row below. If tails turns up, put a tally mark in the T row. Make 20 tosses, then fill in the totals.

Side	Tally	Total
H		
T		

Which side came up more often? _____ Which side do you think will turn up more often if you make 30 more flips? _____

Now flip the coin 30 more times and record your results below.

Side	Tally	Total
H		
T		

Which side came up more often this time? _____ Look at the totals from both sets of flips. If you add all the heads together and all the tails together, which turns up more often? _____

Now exchange coins with a neighbor or your teacher and flip the new coin 30 times. Record your results below.

Side	Tally	Total
H		
T		

Which side came up more often this time? _____

If you have time, make another chart and try someone else's coin. In looking over all your results, does it appear that one side comes up more often than the other? _____

Activity 2   Name _____

What happens when you flip 2 coins at the same time? Try it and see. As you complete each flip of the 2 coins, tally your result below. Do it 40 times.

The Coins Fall	Tally	Total
H,H		
H,T or T,H		
T,T		

1. Which way did the coins fall most often? _____

2. Why do you think they fell this way most often? _____

_____

3. Was there much difference in the number of times you got H,H and the number of

    times you got T,T? _____

4. How do you explain this? _____

_____

5. Enter below the results you might expect to get if you tossed these 2 coins 4 000 times.

The Coins Fall	Total
H,H	
H,T or T,H	
T,T	

Activity 1   Name _____

Throw the die. What number faces up? _____ Throw it again. What number came up this time? _____ What other numbers could come up? _____

EXPERIMENT 1: Try to find out whether one number turns up on your die more often than any other number. Do you think one will? _____ If so, which one do you expect to turn up most often? _____

The table lists all the ways a die can turn up. Every time you throw the die, tally the number that turns up in the appropriate row. Keep throwing the die until one number has 10 tally marks after it. Then find your total for each number.

Face of Die	Tally	Total
⚀		
⚁		
⚂		
⚃		
⚄		
⚅		

EXPERIMENT 2: Let's do the same thing again. Toss the die until one number comes up 10 times. Do you think it will be the same number as before? _____

Face of Die	Tally	Total
⚀		
⚁		
⚂		
⚃		
⚄		
⚅		

[*Continued*]

Activity 1—*Continued*

Name _____

EXPERIMENT 3: Exchange dice with your neighbor or your teacher and see what results you get with the new die. Do this experiment as you did the others, tossing the die until one number has come up 10 times.

Face of Die	Tally	Total
⚀		
⚁		
⚂		
⚃		
⚄		
⚅		

Now take the information from each of your 3 charts and write the totals in the chart below. Then get the grand totals by adding the 3 totals for each face.

Face of Die	Total#1	Total#2	Total#3	Grand Total (#1+ #2+ #3)
⚀				
⚁				
⚂				
⚃				
⚄				
⚅				

Use the table you have just filled in to answer these questions.

1. Did each of the 6 numbers come up at least once in each experiment? _____

2. Look at the grand totals. Did any number turn up twice as often as another one? _____

3. If you were to throw your die once more, would you know what number to expect to turn up? _____

Activity 2   Name _____

Shake the dice and roll them. What sum do you get on the two dice? _____ Shake them again and roll them. What sum did you get this time? _____ Roll the dice 100 times. Record your sums in the box below. When the same sum comes up more than once, use tally marks to show the number of times it comes up. Keep track of your throws by marking off, in the block on the right, each number as you throw. Put a line through one way for the first 50 throws and the other way for the second 50 throws. For example, on toss 1 do this: ⁄; on toss 51 do this: ✗.

Sums on Dice			Number of Throws									
			1	2	3	4	5	6	7	8	9	10
			11	12	13	14	15	16	17	18	19	20
			21	22	23	24	25	26	27	28	29	30
			31	32	33	34	35	36	37	38	39	40
			41	42	43	44	45	46	47	48	49	50

In the chart below, arrange your sums from smallest to largest. After each sum put the total number of times it came up in your experiment. Did you miss any possible sums in your first 100 tosses? If so, put them in their proper place in the chart and record the number of times as 0.

Sums	Times Found in 100 Tosses	Tally	New Total
Smallest:			
Largest:			

[Continued]

Activity 2—*Continued*

Name _____

Look back at your chart for this activity. What sum(s) did you get most often?
_____ How many times did you get it (them)? _____
What sum(s) did you get least often or not at all? _____

Now let's try something a little different. Try to throw sums of five:

or

Keep throwing the dice until you have thrown a sum of five 10 times. In the tally column
in the chart keep a record of all the sums that turn up while you are trying to throw 10 sums
of five.

Now look at the tally column. Which sum has the most tally marks after it? _____
Which sum has the fewest tally marks after it? _____ What sum do you think is your
"lucky sum"? _____ How many times do you think you'd have to throw to get it?
_____ Try. How many throws did it take? _____ Do you think it would take that
long to get it again? _____ Try it and see! How many throws did it take? _____

Activity 3   Name _____

As you have seen, some sums seem to occur more often than others. Study your table from Activity 2. Use the first column of the table below to arrange the sums in order from those you found most often to those you found least often. Then throw the dice another 50 times and tally your results in the middle column. (Do more throws if you have time.) Then find your total for each sum.

Sum on Dice	Tally	Total
Most often:		
Least often:		

Do your sums seem to be in the right order? _____ What changes would you make in the order? _____

Do you think that if you threw your dice a few more times, your results might change? _____

Look at the final totals in your tables for Activities 2 and 3 and then answer these questions.

1. Which sum occurred most often? _____

2. Which sum occurred least often? _____

3. What sums seemed to come up about the same number of times? _____

4. The next time you roll a pair of dice, are you more likely to get a sum of 3 or a sum of 8? _____ Why? _____

_____

Activity 1   Name _____

Draw 3 beads out of your can. What colors are they? _____
Put them back in, stir the beads around with your finger, and draw 3 more. What colors
are they? _____ Let's use letters to stand for the colors.
If you get 2 reds and 1 green, write **RRG**. If you get a red, a blue, and a green, write **RBG**.
Continue drawing 3 beads at a time without looking into the can. Record below the colors
you get each time. Then put the beads back into the can, stir them up, and draw 3 more.
Make 20 draws.

1. _____	6. _____	11. _____	16. _____
2. _____	7. _____	12. _____	17. _____
3. _____	8. _____	13. _____	18. _____
4. _____	9. _____	14. _____	19. _____
5. _____	10. _____	15. _____	20. _____

Now make a list of all the *different* color combinations you have gotten so far. Also write
down the number of times you got each combination.

Color Combination	Tally of Times You Got the Combination	Total

Now make another 20 draws using the same method and record your findings below.

21. _____	26. _____	31. _____	36. _____
22. _____	27. _____	32. _____	37. _____
23. _____	28. _____	33. _____	38. _____
24. _____	29. _____	34. _____	39. _____
25. _____	30. _____	35. _____	40. _____

[*Continued*]

Activity 1—*Continued*   Name _____

Did any combinations come up that you didn't find in your first 20 drawings? _____
If so, what are they? _____ Add these to the chart on the first page of this
activity. Also put in tally marks, so that you can see how many times you got each
combination.

Make 20 more draws and record them below.

41. _____   46. _____   51. _____   56. _____
42. _____   47. _____   52. _____   57. _____
43. _____   48. _____   53. _____   58. _____
44. _____   49. _____   54. _____   59. _____
45. _____   50. _____   55. _____   60. _____

Did you get any new combinations this time? _____ If so, what are they? _____
_____ Add them to your chart and place tally marks after each one for
as many times as you got it in draws 41–60.

Use your chart to answer these questions.

1. What color combination(s) did you draw most often? _____

2. What color combination(s) did you draw least often? _____

Answer these questions after you compare your findings with those of your classmates.

1. Did you draw all the possible combinations? _____

2. If not, which ones are you missing? _____

3. If you were to draw 3 beads at random from the can, what color(s) would you expect

   them to be? _____

Activity 2   Name _____

                                      Name _____

The can contains 36 beads. There is a different number of beads for each color. For example, there could be 18 red, 10 blue, and 8 green beads. Take turns making trial drawings and try to decide the color of the greatest number of beads and the color of the least number of beads. (No fair counting first!)

Draw 3 beads. What colors are they? _____ Put them back in. Do you think that one drawing is enough to let you decide the color of the greatest number of beads? _____

Make 10 draws of 3 beads each, returning the beads before each new drawing. List the color combinations that you get.

1. _____   3. _____   5. _____   7. _____   9. _____

2. _____   4. _____   6. _____   8. _____   10. _____

What color do you think most of the beads are? _____

CONCLUSION 1: There are fewer _____ beads and more _____
beads than any other color.   (color)   (color)

Make 10 more draws of 3 beads at a time and see if these results make you want to change Conclusion 1.

1. _____   3. _____   5. _____   7. _____   9. _____

2. _____   4. _____   6. _____   8. _____   10. _____

[Continued]

Activity 2—*Continued*   Name _____

                                          Name _____

Look at your results so far. Then complete Conclusion 2.

CONCLUSION 2: We think that there are fewer _____ beads and more _____ beads than any other color.

Now make one more check.

1. _____   3. _____   5. _____   7. _____   9. _____

2. _____   4. _____   6. _____   8. _____   10. _____

CONCLUSION 3: Looking at all 30 drawings, we think there are fewer _____ beads and more _____ beads than any other color.

If Conclusions 1, 2, and 3 agree, or if 2 and 3 agree, record your Final Decision below and ask your teacher for another can to decode. If Conclusions 2 and 3 do not agree, make at least 20 more drawings before making your final decision.

1. _____   5. _____   9. _____   13. _____   17. _____

2. _____   6. _____   10. _____   14. _____   18. _____

3. _____   7. _____   11. _____   15. _____   19. _____

4. _____   8. _____   12. _____   16. _____   20. _____

FINAL DECISION: The color of the greatest number of beads is _____.

The color of the fewest number of beads is _____.

(We used the can labeled _____ to make our decision.)

Activity 1   Name _____

Which way do you think a paper or plastic cup will land if you just toss it into the air and let it fall? Put an X under the position you think it will land in most often.

_____   _____   _____

Put a 0 under the way you think it will land least often.

Now toss your cup a few times and see what happens. Make a tally mark in one of the columns to record the way it lands each time. Toss the cup 20 times.

Bottom	Top	Side

Would you like to make a prediction now? If so, circle your prediction.

Most Likely   **Least Likely**

PREDICTION 1:

Make another 20 trials and record your results here.

Bottom	Top	Side

[*Continued*]

Activity 1—*Continued*   Name _____

Study the information from both experiments and then make your second prediction.

Most Likely   **Least Likely**

PREDICTION 2:

Make some more tosses. Make at least 20, but more if you have time.

Bottom	Top	Side

FINAL PREDICTION: I have made _____ tosses of the cup.

On the basis of these tosses, I have decided the following.

Most Likely   **Least Likely**

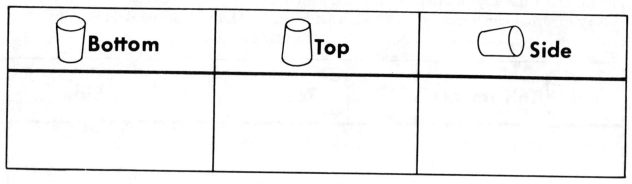

Activity 2   Name _____

Have you ever tossed a thumbtack in the air and had it land point down, sticking into the floor? _____ It doesn't happen very often, does it? How do you think a thumbtack will usually land, like this ⟨🖈⟩ or like this ⟨🖈⟩ ? Circle one.

EXPERIMENT 1: Toss the tack 25 times. Show the way that it lands each time by placing a tally mark in the correct column of row 1.

Experiment	🖈 Lands Up	🖈 Lands on Side	🖈 Lands Down
1			
2			
3			

Which way did it land most often? _____

EXPERIMENT 2: Make another 25 tosses and record them in row 2. Then look at the results of your first 50 tosses. Which way has the tack landed most often now? _____ Did it come up one way twice as often as any other way?_____ If so, which way? _____ Do you think you can predict the way that this tack will usually land? _____ If so, which way? _____

EXPERIMENT 3: If you didn't make a prediction after Experiment 2, do another 50 tosses and record the results in row 3.
If you decided to make a prediction after Experiment 2, check it by doing the 50 tosses.

FINAL CONCLUSION: I have come to the following conclusion based on my tosses.
The most likely way for a thumbtack to land is:

🖈   🖈   🖈

My data indicate that this thumbtack is no more likely to land one way than another. _____

Would a different thumbtack make any difference? _____